Mathematics for Elementary Teachers via Problem Solving

Student Activity Manual

Joanna O. Masingila
Syracuse University

Frank K. Lester
Indiana University

Anne M. Raymond
Bellarmine University

Prentice Hall
Upper Saddle River, New Jersey 07458

Library of Congress Cataloging-in-Publication Data

Masingila, Joanna O. (Joanna Osborne)
    Mathematics for elementary teachers via problem solving: student activity manual/
Joanna O. Masingila, Frank K. Lester, Anne M. Raymond.
      p. cm.
    Includes bibliographical references and index.
    ISBN 0-13-017345-2
    1. Mathematics—Study and teaching (Elementary) 2. Mathematics—Study and teaching—Activity programs. I. Lester, Frank K. II. Raymond, Anne Miller III. Title.

QA135.6 .M37 2002
372.7—dc21

2001034036

Acquisitions Editor: *Quincy McDonald*
Executive Project Manager/Supplements Editor: *Ann Heath*
Editor in Chief: *Sally Yagan*
Vice President/Director of Production and Manufacturing: *David W. Riccardi*
Executive Managing Editor: *Kathleen Schiaparelli*
Senior Managing Editor: *Linda Mihatov Behrens*
Production Editor: *Jami Darby, WestWords Inc.*
Production Assistant: *Nancy Bauer*
Manufacturing Buyer: *Alan Fischer*
Manufacturing Manager: *Trudy Pisciotti*
Marketing Manager: *Patrice Lumumba Jones*
Assistant Editor of Media: *Vince Jansen*
Editorial Assistant: *Joanne Wendelken*
Art Director: *Maureen Eide*
Assistant to the Art Director: *John Christiana*
Interior Design: *Jill Little*
Cover Design: *Joseph Sengotta*
Managing Editor, Audio/Video Assets: *Grace Hazeldine*
Creative Director: *Carole Anson*
Director of Creatice Services: *Paul Belfanti*
Cover Photo: *Scott Cunningham, PH Merrill Publishing*
Art Studio: *Network Graphics*

 © 2002 by Prentice-Hall, Inc.
Upper Saddle River, New Jersey 07458

All rights reserved. No part of this book may be reproduced, in any form or by any means, without permission in writing from the publisher.

Printed in the United States of America
10 9 8 7 6 5 4 3 2 1

ISBN 0-13-017345-2

Pearson Education LTD., *London*
Pearson Education PTY, Limited, Australia *Sydney*
Pearson Education North Asia Ltd, *Hong Kong*
Pearson Education Canada, Ltd., *Toronto*
Pearson Educación de Mexico S.A. de C.V.
Pearson Education—Japan, *Tokyo*
Pearson Education Malaysia, Pte. Ltd.

DEDICATION

To Frank Lester, who opened my eyes to teaching and learning mathematics via problem solving.

JOM

To Sandy Kerr and John LeBlanc, who made me start thinking differently about what teaching and learning mathematics involves.

FKL

To my family.

AMR

Brief Contents

Chapter 1: Getting Started in Learning Mathematics via Problem Solving 1

Chapter 2: Numeration 39

Chapter 3: Operations on Natural Numbers, Whole Numbers, & Integers 71

Chapter 4: Number Theory 109

Chapter 5: Data & Chance 151

Chapter 6: Fraction Models & Operations 223

Chapter 7: Real Numbers: Rationals & Irrationals 257

Chapter 8: Patterns & Functions 287

Chapter 9: Geometry 355

Chapter 10: Measurement 409

Appendix A: Pages to Accompany Selected Activities 473

Answers to Odd-numbered Selected Exercises and More Problems 483

Contents

Preface xv
References xix

Chapter 1: Getting Started in Learning Mathematics via Problem Solving 1

Activity 1.1   What's My Number?   2

Activity 1.2   Poison   4

Activity 1.3   Cereal Boxes and Patio Tiles   6

*Activity 1.4   The Mathematics in the Pages of a Newspaper   9

Activity 1.5   Family Relations   13

Activity 1.6   Constructing Numbers   15

Activity 1.7   Problem-Solving Tips   17

Activity 1.8   The Valentine's Day Party   24

Activity 1.9   The Puzzle of the Hefty Hippos   25

Activity 1.10   Making Dice   26

Activity 1.11   The Tower of Hanoi   28

Things to Know   30

Exercises & More Problems   30

Chapter 2: Numeration 39

Numeration Systems

Activity 2.1   Early Numeration Systems   40

Activity 2.2   The Hindu-Arabic Numeration System   42

Activity 2.3   Comparing Numeration Systems   45

Activity 2.4   Mathematical and Nonmathematical Characteristics of Systems   47

Activity 2.5   Creating a Numeration System   48

Understanding Place Value Through Different Bases

Activity 2.6   Exploring Place Value Through Trading Games   49

Activity 2.7   Converting from One Base to Another   52

Activity 2.8   Place Value and Different Bases   54

*Activity 2.9   Solving a Problem with a Different Base   56

Activities marked with an * are designed to be worked using technology tool. See the Preface or activity for more information.

Activity 2.10   Computations in Different Bases   57

Things to Know   59

Exercises & More Problems   60

Chapter 3: Operations on Natural Numbers, Whole Numbers, & Integers   71

Natural Numbers, Whole Numbers, Integers, and Their Properties

Activity 3.1   Exploring Sets of Numbers   72

Activity 3.2   Sets and Their Properties   78

Operations with Whole Numbers

Activity 3.3   Classifying Word Problems by Operation   81

Operations with Integers

Activity 3.4   Integer Addition and Subtraction   83

Activity 3.5   Integer Multiplication and Division   87

Algorithms

Activity 3.6   Scratch Addition Algorithm   91

Activity 3.7   Lattice Multiplication Algorithm   92

Activity 3.8   Cashiers' Algorithm   93

Activity 3.9   Austrian Subtraction Algorithm   94

Activity 3.10   Russian Peasant Algorithm   95

Activity 3.11   Using Algorithms to Solve Problems   96

*Activity 3.12   Operation Applications   98

Things to Know   101

Exercises & More Problems   101

Chapter 4: Number Theory   109

Primes, Composites, and Prime Factorization

Activity 4.1   The Locker Problem   110

Activity 4.2   Searching for Patterns of Factors   111

Activity 4.3   Factor Feat: A Game of Factors   113

Activity 4.4   Classifying Numbers According to Prime Factorization   114

Activity 4.5   E-Primes   119

Activity 4.6   Twin Primes and Prime Triples   121

Divisibility

Activity 4.7   Divisibility Tests   123

Activity 4.8   Divisibility in Different Bases   124

**Activity 4.9*   Factors and Multiples   125

Modular Arithmetic

Activity 4.10   A Different Way of Counting   127

Activity 4.11   Operations with Modular Arithmetic   128

Activity 4.12   Mystery Numeration System   129

Representations of Number-Theory Ideas

Activity 4.13   Figurate Numbers   130

Activity 4.14   Number Ideas: Proofs Without Words   132

Activity 4.15   The Fibonacci Sequence   133

Activity 4.16   Pascal's Triangle   137

Things to Know   139

Exercises & More Problems   139

Chapter 5:   Data & Chance   151

Probability

Activity 5.1   Two Probability Experiments: Spinners & Color Tiles   152

Activity 5.2   Probability Experiments with Dice & Chips   158

Activity 5.3   Are These Dice Games Fair?   161

Activity 5.4   What Would Marilyn Say?   164

Activity 5.5   Basic Probability Notions   165

Activity 5.6   Probability Models of Real-World Situations   168

**Activity 5.7*   Basic Counting Principles   171

Statistics

**Activity 5.8*   Using Statistics to Summarize Data   179

**Activity 5.9*   Using Statistics in Decision Making   186

**Activity 5.10*   Looking at Variability in Data: Part I   191

Activity 5.11   Looking at Variability in Data: Part II   196

**Activity 5.12*   Chirping Crickets and Temperature: A Correlation Problem   199

Activity 5.13   Statistics and Sampling   203

Things to Know   210

Exercises & More Problems   211

Chapter 6: Fraction Models & Operations 223

Three Ways to Represent Fractions

Activity 6.1 Introducing the Region Model 224

Activity 6.2 Introducing the Linear Model 225

Activity 6.3 Introducing the Set Model 226

Using the Region Model of Fractions

Activity 6.4 Exploring Fraction Ideas Through the Region Model 227

Activity 6.5 Fractions on the Square: A Game Using the Region Model 229

Activity 6.6 Fraction Puzzles Using the Region Model 230

Using the Linear Model of Fractions

Activity 6.7 Looking for Patterns with the Linear Model 231

Activity 6.8 Exploring Fraction Ideas Through the Linear Model 232

Activity 6.9 Exploring the Density of the Set of Real Numbers 235

Using the Set Model of Fractions

Activity 6.10 Exploring Fraction Ideas Through the Set Model 237

Activity 6.11 Solving Problems Using the Set Model 239

Operations with Fractions

Activity 6.12 Classifying Problems by Operation: Revisiting Activity 3.3 240

Activity 6.13 Illustrating Operations with Region, Linear, and Set Models 241

Activity 6.14 Developing Fraction Sense with Linear Models 245

Activity 6.15 Using the Region Model to Illustrate Multiplication 247

Activity 6.16 Using the Region Model to Illustrate Division 248

Things to Know 250

Exercises & More Problems 250

Chapter 7: Real Numbers: Rationals & Irrationals 257

Ratio and Proportion

Activity 7.1 Exploring Ratio and Proportion Ideas 258

*****Activity 7.2** Solving Problems Using Proportions 259

*****Activity 7.3** Graphing Proportion Problems 261

A Fourth Way to Represent Fractions

Activity 7.4 Introducing Decimal Representation 262

Activity 7.5 Explaining Decimal-Point Placement 264

Activity 7.6   Modeling Operations with Decimals   266

Activity 7.7   Converting Decimals to Fractions   267

A Fifth Way to Represent Fractions

*Activity 7.8   Introducing Percent Representation   270

*Activity 7.9   Pay Those Taxes: A Game of Percents and Primes   272

*Activity 7.10   Comparing Fractions, Decimals, and Percents   273

*Activity 7.11   Four in a Row: A Game of Decimals and Factors   274

Exploring Irrational Numbers

Activity 7.12   Exploring Circles: Approximating an Irrational Number   275

Activity 7.13   Constructing Irrational Numbers   277

Activity 7.14   Properties of Rational and Irrational Numbers   279

Things to Know   281

Exercises & More Problems   281

Chapter 8: Patterns & Functions   287

Variables

*Activity 8.1   Exploring Variables   288

*Activity 8.2   Exploring Variables Through Data in Tables   291

*Activity 8.3   Exploring Variables Through Data in Graphs   295

Activity 8.4   Interpreting Graphs   299

Patterns

*Activity 8.5   Investigating and Describing Numerical Patterns   303

*Activity 8.6   Investigating Numerical Situations   305

*Activity 8.7   Identifying Rules and Functions   308

*Activity 8.8   Looking for an Optimal Solution   311

Iteration

*Activity 8.9   Investigating Numerical Functions That Repeat   317

*Activity 8.10   Investigating Real-Life Iteration: Savings-Account Interest   320

*Activity 8.11   Investigating Iteration: Geometry and Fractals   323

Functions and Equations

Activity 8.12   Properties of Equations   327

Activity 8.13   Using Properties to Solve Equations   329

*Activity 8.14   Investigating Distance vs. Time Motion   330

xii   Contents

Activity 8.15   **Constructing and Interpreting Sensible Graphs**   336

Things to Know   341

Exercises & More Problems   341

Chapter 9: Geometry   355

Thinking Mathematically and Geometrically

Activity 9.1   **Communicating with Precise Language**   356

Activity 9.2   **Definitions: What is Necessary, and What is Sufficient?**   357

Activity 9.3   **Tangram Puzzles: Exploring Geometric Shape**   358

Lines and Angles

Activity 9.4   **Constructing Geometric Relationships**   360

Activity 9.5   **Exploring Lines and Angles**   362

Activity 9.6   **Defining Angles and Lines**   367

Polygons

Activity 9.7   **Exploring Polygons**   371

Activity 9.8   **Exploring Quadrilaterals**   373

Activity 9.9*   **Properties of Quadrilaterals   375

Activity 9.10   **Defining Triangles**   377

Activity 9.11   **Exploring Side Lengths in Triangles**   381

Congruence and Similarity

Activity 9.12   **Exploring Triangle Congruence**   382

Activity 9.13   **Exploring Triangle Similarity**   385

Activity 9.14   **More on Similarity**   388

Proof

Activity 9.15   **Constructing Proofs**   389

Activity 9.16   **Sums of Measures of Angles of Polygons**   391

Exploring Non-Euclidean Geometry

Activity 9.17   **Spherical Geometry**   393

Things to Know   395

Exercises & More Problems   396

Chapter 10: Measurement   409

One-and Two-dimensional Measurement

Activity 10.1   Exploring Area and Perimeter   410

Activity 10.2   Perimeter and Area: Is There a Relationship?   412

Activity 10.3   Pick's Formula   415

Activity 10.4   Investigating Length   416

Activity 10.5   Pythagoras and Proof   419

Activity 10.6   Investigating Circles   426

Activity 10.7   Investigating the Circumference-to-Diameter Ratio   429

Activity 10.8   Investigating the Area of a Circle   431

Three-dimensional Measurement

Activity 10.9   Surface Area and Volume of Rectangular Prisms   435

Activity 10.10   Drawing Rectangular Prisms   437

Activity 10.11   Exploring the Surface Area of Cones   440

Activity 10.12   Investigating the Volumes of Cylinders and Cones   442

Analytic and Transformational Geometry

Activity 10.13   Geoboard Battleship: Exploring Coordinate Geometry   446

*Activity 10.14   Investigating Translations using the Geometer's Sketchpad ™   449

*Activity 10.15   Investigating Rotations using the Geometer's Sketchpad ™   452

*Activity 10.16   Investigating Reflections using the Geometer's Sketchpad ™   454

Tessellations

Activity 10.17   Tessellations: One Definition   456

Activity 10.18   Tessellations: Another Definition   459

Things to Know   461

Exercises & More Problems   462

*Appendix A: Pages to Accompany Selected Activities   473

Answers to Odd-numbered Exercises and More Problems   483

Preface

Give a child a fish and you feed her for a day. Teach a child to fish and you feed her for life.

The ancient Chinese saying above tells us how important it is to have good teachers for our children. Indeed, there are few professions or occupations that are as important to the welfare of our society and culture as teaching. The purpose of the activities in this book is to help you develop a deep and lasting understanding of the mathematical concepts, procedures, and skills that are essential to being able to teach young children, in particular children in the elementary grades. We believe that if you develop such deep and lasting understanding, you will be well prepared to teach mathematics to many children and thereby help to prepare them to lead productive, informed lives once their school days are over.

The Way Mathematics Is Taught Is Changing

Those who would teach mathematics need to learn contemporary mathematics appropriate to the grades they will teach, in a style consistent with the way in which they will be expected to teach.

All students, and especially prospective teachers, should learn mathematics as a process of constructing and interpreting patterns, of discovering strategies for solving problems, and of exploring the beauty and applications of mathematics.
(*Everybody Counts: A Report to the Nation on the Future of Mathematics Education*, 1989, pp. 64, 66)

These two quotes shown are taken from a report written over 10 years ago for the United States' National Research Council by a group of concerned mathematics teachers. The authors of the report insisted that it was time to change the way that mathematics was taught at all levels, kindergarten through university. Since this report was written, the nature of mathematics instruction has begun to change. In the past, mathematics instruction was viewed by many as an activity in which an "expert"—usually the teacher—attempted to transmit her or his knowledge of mathematics to a group of students who usually sat quietly trying to make sense of what the expert was telling them. This passive transmission view has been replaced by a new view in which mathematics is seen as a cooperative venture among students who are encouraged to explore, make and debate conjectures, build connections among concepts, solve problems growing out of their explorations, and construct personal meaning from all of these experiences.

Principles of the Problem-Based Approach

The activities contained in our books have been created with the new view of mathematics teaching and learning promoted by the *American Mathematical Association of Two-Year Colleges* (AMATYC) and the *National Council of Teachers of Mathematics* (NCTM). In particular, we developed the activities with the following documents in mind: AMATYC's publication, *Crossroads in Mathematics: Standards for Introductory College Mathematics Before Calculus*, and the NCTM's publications, *Curriculum and Evaluation Standards for School Mathematics, Professional Standards for Teaching Mathematics, Assessment Standards for School Mathematics*, and *Principles and Standards for School Mathematics*.

From these five documents we developed a set of principles to guide the development of all activities.

1. *All Activities Are Based on the NCTM Standards.*

Special emphasis is placed on the five process standards of the NCTM: *problem solving, communication, reasoning, connections*, and *representations*. First and foremost, students should be engaged in the solution of thought provoking problems. Not only should students learn to solve problems, but they should also learn mathematics *via* problem solving. The second major standard is *communication*. Knowing mathematics is of little value if one cannot communicate mathematical ideas to other people. NCTM's third major standard is *reasoning*. Among other things, reasoning deals with the ability to think through a problem and to carefully evaluate any solution that has been proposed. The fourth of the major standards involves making *connections*. To really understand mathematics, one must be able to see connections between various mathematical ideas, and between "school" and "real world" mathematics. Finally, the way in which mathematical ideas are represented is vital to how students can understand and apply those ideas. Representations should be viewed as essential ingredients in supporting the development of deep understanding.

2. Solving Problems Regularly and Often Is an Essential Part of Developing a Good Understanding of Mathematics.

In order for you to improve your ability to solve mathematics problems, you must attempt to solve a variety of types of problems on a regular basis and over a prolonged period of time. We also believe that ability to solve problems goes hand-in-hand with the development of an understanding of mathematical concepts, procedures, and skills. Put another way, as you solve problems you will develop better understanding of the mathematics involved in the problems. And, as you develop better understanding of mathematical ideas, you will become a better problem solver.

3. Problem Solving Involves a Very Complex Set of Processes.

There is a dynamic interaction between mathematical concepts and the processes used to solve problems involving those concepts. That is, heuristics, procedural skills, control processes, awareness of one's cognitive processes, etc. develop concurrently with the development of an understanding of mathematical concepts.

4. The Teacher's Role in Fostering Healthy Problem-solving Performance Is Vitally Important.

Problem-solving instruction is likely to be most effective when it is provided in a systematically organized manner under the direction of the teacher. Our philosophy is that the role of the teacher changes from that of a "dispenser of knowledge" to a "facilitator of learning." With respect to problem solving and reasoning, this implies that the teacher does very little lecturing on how to solve specific types of problems and much more posing and discussing of a wide variety of non-routine and applied problems. The teacher also focuses on helping you make connections between the mathematics you are learning and its application to the workplace or home.

5. Cooperative, Small-group Work Is Encouraged.

The standard arrangement for working on the activities in the *Student Activity Manual* is for you to work in small groups. Small group work is especially appropriate for activities involving new content (e.g., new mathematics topics, new problem-solving-strategies) or when the focus of the activity is on the process of solving problems (e.g., planning, decision making, assessing progress) or exploring mathematical ideas.

6. Assessment Practices Are Closely Connected to Instructional Emphases.

We believe that the teacher's instructional plan should include attention to how your performance will be assessed. In order for you to become convinced of the importance of the sort of behaviors that a good problem-solving program promotes, it is necessary to use assessment techniques that reward such behaviors. As a result, we encourage teachers to use various alternative assessment methods such as providing opportunities during tests for you to work with a group of your classmates to solve certain problems on the tests. We also encourage teachers to assess your ability to discuss your understanding of mathematical concepts and procedures in writing and orally.

Features

Hands-on Exploration through Group Work

The activities in this text are designed to engage you in doing real mathematics through small-group exploration. We have two mottoes that should be followed:

- Mathematics is best learned by active, "hands-on" exploration of real problems.
- If "two heads are better than one," then three or four heads are even better!

These mottoes arise from our conviction that the best learning occurs when you are engaged actively in making sense out of problematic situations. Thus, it is your responsibility to make sense out of the activities, rather than wait for the teacher to tell you what is important or how to solve the problems.

Consequently, the activities in this book include almost no explanations with them. It will be your responsibility to work with the students in your group (some students like to refer to their groups as "teams") to solve problems and develop good understanding of the mathematics involved. The teacher's job is to encourage you, to offer gentle assistance without giving too much specific guidance. This style of learning may be a new experience for you and it may even be a bit uncomfortable for you at first. But, be patient! As you gain experience working in a group with others rather than depending on the teacher to tell you everything you should know, you are likely to find that you are becoming more and more independent of the teacher and increasingly in control of your own learning.

Activities Grouped by Chapter

The activities are broken into chapters that conform to ten different mathematical topics. Each chapter begins with

- an overview of the subject addressed in the activities and an outline of the activities
- "The Big Mathematical Ideas" addressed in the activities
- "NCTM Principles and Standards Links"

Each chapter concludes with "Things to Know" that reviews your understanding of the material and includes lists of

- Words to Know
- Concepts to Know

A set of "Exercises & More Problems" helps test your mastery of the skills learned in the activities and expand your grasp of the concepts. This section is divided into several different types of problem sets including:

- Exercises
- Critical Thinking
- Extending the Activity
- Writing and Discussing

Activities Linked to Student Resource Handbook

As noted previously, the activities in this manual do not include text discussions and explanations of concepts. To provide this information with the activities would be detrimental to helping you become a good problem solver and an independent learner of mathematics. However, we developed the *Student Resource Handbook* to provide you with useful explanations of the mathematical concepts and procedures that are explored in the activities. *The Student Resource Handbook* and *Student Activity Manual* are designed to work hand-in-glove.

- **FYI** At the beginning of each activity, an FYI box lists relevant topics in *The Student Resource Handbook* that correspond to the content of the activity. You may find it helpful to review the referenced topic as a content refresher while working the activity.
- **Tools Appendix**—A selection of the paper manipulatives used in selected activities are found in the Tools Appendix of the *Student Resource Handbook*. You may find it helpful to make copies of those tools in preparation for working a given activity. In addition, the Tools Appendix includes lists of useful resources for teachers and a synopsis of the NCTM Standards.

Use of Technology

The use of technology tools ranging from a Graphing Calculator, to a statistical software package, to Geometer's Sketchpad can be valuable in exploring mathematical ideas. Although we do not provide any instruction on how to use various tools, a number of our activities are written with the assumption that some sort of technology tool will be used while working them. These activities are denoted with an asterisk in the preface and are listed on the following page by the technology assumption:

Technology Tool	Activity
Calculator	1.4, 1.9, 2.9, 3.12, 4.9, 5.7, 7.2, 7.8, 7.9, 7.10, 7.11, 8.2, 8.5, 8.7, 8.8, 8.9, 8.10
Graphing Calculator With CBL or CBR unit	5.8, 5.9, 5.10, 5.12, 7.3, 8.1, 8.3, 8.6, 8.14
Statistical Software package	5.8, 5.9, 5.10, 5.12
Spreadsheet	8.6
Geometer's Sketchpad	9.9, 10.14, 10.15, 10.16

Supplements

Web Site (*www.prenhall.com/Masingila*)
The Web site includes a number of support features for students and faculty including:

- **"Projects"**–a selection of more complex group activities that take more time than the activities in the text and often require outside research be done to arrive at a solution.
- Links to useful articles and publications that suggest scoring rubrics for exploratory, group based classes, teaching suggestions, etc.
- Contact information for the authors to provide a support for instructors as they embark on this new approach to teaching.
- Bulletin Boards for shared information

Instructor's Resource Manual (ISBN 0-13-018989-8)- a complete set of solutions to the activities in the text is available from the publisher to adopters of the *Mathematics for Elementary Teachers via Problem Solving*.

Acknowledgments

Although it is impossible to acknowledge all of the people who have helped us in writing this textbook, we want to thank some people specifically. Since these materials were developed out of two projects funded by the National Science Foundation (one at Indiana University and one at Syracuse University), there have been a number of people who developed activities and ideas for activities that have contributed to this textbook set. We are grateful to these people and want their contributions recognized here: Jean-Marc Cenet, Rapti de Silva, K. Jamie King, Norman Krumpe, Diana V. Lambdin, Sue Tinsley Mau, Francisco Egger Moellwald, Preety Nigam, and Vânia Santos. There were also people who helped in the latter stages of writing the preliminary and first editions by generating exercises and solutions, checking for accuracy, and developing particular sections of the Student Resource Manual. We thank them for their help: Fran Arbaugh, Zaur Berkaliev, Ernesto Colunga, Rapti de Silva, Dasha Kinelovsky, Levi Molenje, Jean Palm, Sandra Reynolds, and Robert Wenta.

We also thank the following individuals who reviewed our manuscript. Their comments and suggestions have helped us make this a better textbook set: Rita M. Basta (*California State University at Northridge*), George Csordas, (*University of Hawaii at Mànoa*), Richard Friedlander (*University of Missouri at St. Louis*), Michael Hall (*University of Mississippi at Oxford*), Guershon Harel (*University of California at San Diego*), Ted Hodgson (*Montana State University*), Donald Hooley (*Bluffton College*), Eric Milou (*Rowan State University*), Kathy Nickell (*College of Dupage*), Dale Oliver, (*Humboldt State University*). Finally, we thank our editors, Ann Heath, Quincy McDonald, and Sally Yagan, for their encouragement and desire to publish a reform textbook for prospective elementary teachers.

We have worked hard to make sure that our books are as clean and accurate as possible. However, if you identify any errors in the text, please send us the information so that it may be corrected in subsequent printings of the text and posted on our web site.

Joanna O. Masingila
(jomasing@syr.edu)

Frank K. Lester
(lester@indiana.edu)

Anne M. Raymond
(araymond@bellarmine.edu)

References

American Mathematical Association of Two-Year Colleges. (1995). *Crossroads in Mathematics: Standards for Introductory College Mathematics Before Calculus*. Memphis, TN: Author.

Mathematical Sciences Education Board (National Research Council). (1989). *Everybody Counts: A Report to the Nation on the Future of Mathematics Education*. Washington, DC: National Academy Press.

National Council of Teachers of Mathematics. (1989). *Curriculum and Evaluation Standards for School Mathematics*. Reston, VA: Author.

National Council of Teachers of Mathematics. (1991). *Professional Standards for Teaching Mathematics*. Reston, VA: Author.

National Council of Teachers of Mathematics. (1995). *Assessment Standards for School Mathematics*. Reston, VA: Author.

National Council of Teachers of Mathematics. (2000). *Principles and Standards for School Mathematics*. Reston, VA: Author.

CHAPTER ONE

Getting Started in Learning Mathematics via Problem Solving

CHAPTER OVERVIEW

In this chapter, you will begin to see just how different this course is from any other mathematics course you have taken. The activities in this chapter focus on three things: (1) learning how to work *in a small, cooperative group* on real mathematical investigations, (2) becoming *less dependent on your instructor* for answers and direction, and (3) learning about certain *key problem-solving strategies*. All of this will be done in the context of playing various strategy games and exploring solutions to some interesting problems. Throughout the activities, you will be gaining valuable experience in how to collaborate in a productive way with others without relying on your instructor. You will also be looking for patterns, guessing and checking, making conjectures, using logical reasoning, and making organized lists.

BIG MATHEMATICAL IDEAS

Problem-solving strategies, generalizing, verifying, using language and symbolism, multiple representations

NCTM PRINCIPLES & STANDARDS LINKS

Problem Solving; Reasoning; Communication; Connections; Representation

Activity **1.1** What's My Number?
1.2 Poison
1.3 Cereal Boxes and Patio Tiles
1.4 The Mathematics in the Newspaper
1.5 Family Relations
1.6 Constructing Numbers
1.7 Problem-Solving Tips
1.8 The Valentine's Day Party
1.9 The Puzzle of the Hefty Hippos
1.10 Making Dice
1.11 The Tower of Hanoi

Activity 1.1 What's My Number?

FYI Topics in the Student Resource Handbook

1.5 Problem-Solving Topics

In your group, form two teams. One team will play against the other team in the group.

Rules for playing *What's My Number?*
  i. One team, say team A, writes a three-digit number on a piece of paper (e.g., 982).
  ii. All digits in the number must be different.
  iii. The other team, say team B, tries to determine the number that has been written on the paper by naming any three-digit number.
  iv. In return, team A tells team B the number of correct digits and the number of correct places in their guess.
  v. Team B continues to name numbers until it determines the correct number.

Play several games of What's My Number? Alternate which team forms the number and which team guesses the number. You will want to have an organized method of keeping track of your guesses and the information you receive from the other team. Try to determine strategies that minimize the number of guesses you need. See if you can answer the following questions.

1. Was either team able to develop a strategy for playing the game? Describe any strategies the teams used.

2. Which team has the better strategy? What does it mean to have a "better strategy"?

3. A strategy is the best possible strategy if it minimizes the number of numbers guessed before correctly identifying the number. Was either team able to devise the best possible strategy? How do you know?

Activity 1.2 Poison

FYI Topics in the Student Resource Handbook
1.5 **Problem-Solving Topics**

Form two teams in your group. One team will play against the other team. Your instructor will give you 10 color tiles. Place the 10 tiles between the two teams, and follow these rules:

i. Decide which team will go first.
ii. When it is your team's turn, you must take one or two tiles from the table.
iii. Alternate turns until there are no tiles remaining on the table.
iv. The team who takes the last tile, the "poison" one, is the loser.

1. Play several games of *Poison*, and try to determine a good strategy for winning.

2. What if you start a game with nine tiles? With 11 tiles? With 12 tiles?

3. Can you determine a strategy for winning *Poison* no matter how many tiles you have at the start of the game?

4. What makes a strategy a good strategy?

Activity 1.3 Cereal Boxes and Patio Tiles

FYI Topics in the Student Resource Handbook
1.5 Problem-Solving Topics
4.8 Patterns

PROBLEM 1: Part A: Stacking Cereal Boxes

A store clerk was told that she had 45 cereal boxes to be stacked in the display window and that all of the boxes had to be used. The manager told the clerk that the boxes had to be in a triangle, like the one shown below. The sales clerk wondered how many boxes needed to be placed on the bottom row to build the triangle, using all the boxes.

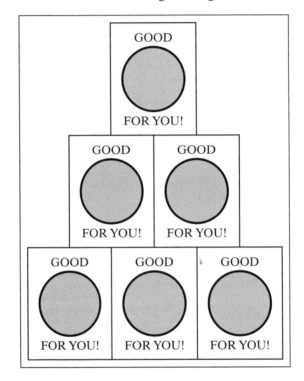

Part B: Stacking More Boxes

Suppose the clerk had to stack 200 boxes of cereal in a triangle, like the one above. Now, how many boxes would be on the bottom row? What if the clerk had B boxes to stack?

PROBLEM 2: Laying Blocks in a Patio

A patio was to be laid in a design like the one shown. A man had 50 blocks to use. How many blocks should be placed in the middle row to use the most number of blocks?

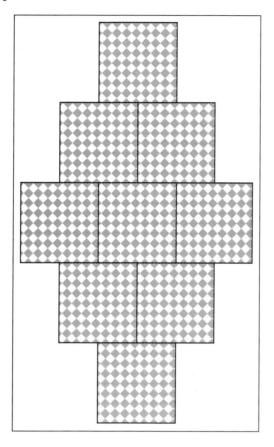

Generalize your answer to the patio-block problem.

PROBLEM 3: Looking for Similarities

How are problems 1 and 2 alike? How are they different?

Assignment: Talk over your solutions with your group. Decide what gave you trouble and why. Ask yourselves questions such as: "What did we do that helped us get started in the right direction? What did we do that gave us trouble? Why?"

Activity 1.4 The Mathematics in the Pages of a Newspaper

FYI Topics in the Student Resource Handbook

1.5 Problem-Solving Topics

Newspapers usually consist of large sheets of paper that are printed and folded in half to form the pages of a newspaper.

1. Suppose you had a sheet of paper that you folded in half. How many pages would be formed with one such sheet of paper? How many pages would be formed with two such sheets?

2. If, starting with 1, you numbered the pages of the single sheet that was folded in half, what would be the numbers that appeared on either side of the sheet? Answer the same question if you had two sheets of paper arranged as in a newspaper.

3. What is the sum of the page numbers on the same side of the sheet for the single sheet? What is this sum for the double sheet?

4. What is the sum of all page numbers for the single sheet? For the double sheet?

5. a. If you had 10 sheets of paper in the newspaper, how many pages would you get?

   b. What would be the page numbers that would appear on the innermost sheet? What would be their sum?

   c. What would be the sum of the page numbers that appeared on any one side of a sheet of paper?

   d. What would be the sum of all the pages in the newspaper?

6. Answer questions #2–5 if you had 100 sheets of paper.

7. a. If your newspaper had to have 28 pages, how many sheets of paper would you need?

   b. What would be the sum of the page numbers that appeared on any one side of a sheet of paper?

c. What would be the page number that appeared on the same side of the sheet as page 8?

8. Answer the above questions if your newspaper had 50 pages.

9. Pat pulled out one full sheet from the Travel section of *The New York Times*. If the left half of the sheet was numbered 4 and the right half was numbered 19, how many sheets were there in the Travel section? Had there been 18 sheets in the section, which page would have appeared opposite the page numbered 6?

Activity 1.5 Family Relations

FYI Topics in the Student Resource Handbook
1.4 Logic
1.5 Problem-Solving Topics

The following is part of a family tree. Answer the questions that follow based on the tree given below. Make sure that you state the relationship as precisely as possible.

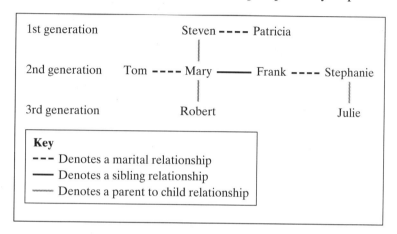

Key
- - - Denotes a marital relationship
——— Denotes a sibling relationship
▬▬▬ Denotes a parent to child relationship

1. How is Patricia related to Frank?

2. How is Steven related to Stephanie?

3. How is Julie related to Patricia?

4. How is Mary related to Stephanie?

5. How is Tom related to Steven?

6. How is Frank related to Tom?

7. How is Tom related to Stephanie?

8. How is Robert related to Tom?

9. How is Julie related to Mary?

10. How is Robert related to Frank?

11. If Stephanie met her husband's sister's father on Tuesday, whom did she meet?

12. If Tom called his wife's brother's daughter, whom did he call?

13. If Frank invited his mother's grandson for dinner, whom would he have invited?

14. If Robert visited his mother's brother's wife yesterday, whom did he visit?

Activity 1.6 Constructing Numbers

FYI Topics in the Student Resource Handbook
1.5 Problem-Solving Topics

Using each of the 10 digits no more than once, construct the numbers described below:

1. largest seven-digit odd number with a 9 in the tens place

2. largest 10-digit even number containing more than 50% odd digits

3. smallest six-digit even number with a 1 in the thousands place

4. largest eight-digit even number

5. smallest nine-digit even number

6. number closest to a half-billion

7. smallest six-digit multiple of 5 containing no digit less than 6

8. number closest to a half-million that contains no digit less than 4

9. largest 10-digit number that is a multiple of 10

10. smallest seven-digit number containing no digit between 1 and 5

11. largest six-digit even number that is a multiple of 5 and uses no digits between 4 and 9

12. smallest 10-digit odd number with more than 75% of the digits greater than 4

13. largest six-digit multiple of 9 that is odd

14. largest odd number using only digits that alone are considered prime numbers

15. number closest to a quarter-billion that contains only even digits (you may repeat digits)

Activity 1.7 Problem-Solving Tips

FYI Topics in the Student Resource Handbook
1.5 Problem-Solving Topics

Perhaps the most important goal of this course is to improve your ability to think and reason mathematically. As you gain in confidence to engage in mathematical thinking and reasoning, you will also notice that you are becoming more able to use the mathematics you know to solve problems. As you continue to work on the problems in this book, you will want to consider a few general guidelines, or tips, for solving math problems. (These tips are listed and discussed on page 20.) Throughout this course, as you work on the activities and problems, either with your group or alone, you will want to refer to these tips. The activity described below is intended to cause you to think seriously about how you thought about and attempted to solve the problems you worked on so far in this course.

Activity Description: Reflecting on Yourself as a Problem Solver

1. Solve the following problem:

   Eight children are entered in a table tennis tournament. If each child plays one game with each of the other children in the tournament, how many games will be played altogether?

2. What did you do, either individually or as a group, to help you understand what the problem is asking and what you needed to do? Why did you do these things?

3. How did you decide how to begin to work on the problem? What gave you the idea for your plan for solving it?

4. Were you sure that your plan for solving the problem was correct? Were you sure you were actually implementing your plan correctly? Explain your answer!

5. When you finished working on the problem and had arrived at a solution, how sure were you that it was correct? What did you do to determine if your solution was reasonable? How did you check your work? When did you check your work (only after you were finished or during your solution effort)?

6. Could you now solve this problem no matter how many children entered the tournament? Are you sure? What if there were n children in the tournament? Could you express the answer no matter how many children were entered? Explain how you could do this?

7. Read the discussion entitled "Why Is It Important to Learn How to Solve Mathematics Problems?" Are any of the approaches for solving the table-tennis problem described in the discussion similar to how you solved it? Which approach is most like the way your group solved this problem?

8. Near the end of the discussion, mention is made of triangular numbers. What would the 25th triangular number be? The 100th?

Tips for Solving Mathematics Problems

I. Develop Good Understanding of the Problem. To be successful in solving any problem, you must understand it well.

Suggestions: (a) Restate the problem in language that is clear and sensible to you; (b) Clarify the question (including any hidden questions); (c) Organize the information (get rid of unneeded information and find needed and assumed information).

II. Devise a Good Plan of Attack on the Problem. Analyze the problem with the goal of identifying a systematic method for solving the problem.

Suggestions: (a) Ask yourself: "Have I ever solved a problem like this one before? In what ways was it like this one? What did I do to solve that problem that might be helpful in solving this one?" (b) Consider the various problem-solving strategies that you have learned—decide if any of them will help; (c) Determine what you want to do first, second, and so on.

III. Carefully Carry Out the Plan. Implement the plan of attack that you have chosen, being careful to make sure that you are implementing it correctly.

Suggestions: (a) Check your work along the way as you proceed with your solution effort; (b) Don't be content to check only your computations. Also, check to make sure that you are using all the important information and that you have not misinterpreted anything; (c) Ask yourself: "Is this plan getting me anywhere? Am I following the plan I chose, or am I becoming sidetracked? Should I abandon this plan and try to think of another way to solve the problem?"

IV. Evaluate Your Solution and Think About What You Have Learned. Decide if the solution is reasonable, and consider what you have learned by solving the problem, not only about mathematics, but also about yourself as a doer of mathematics.

Suggestions: (a) Check your solution with the important information; (b) Check all your computations and other work one more time; (c) Think about the plan you used and ask yourself: "Could I have solved this problem another, better way? Would my plan work if the numbers were larger? Would my plan work in general? Could I now solve other problems similar to this one?"

Why Is It Important to Learn How to Solve Mathematics Problems?*

Why should children study mathematical problem solving? A complete answer to this question is not possible in a few words, but central to an answer is our belief that engaging children in problem-solving activities: (1) provides them with opportunities to extend their thinking processes and (2) encourages them to make mathematical connections among the problems they are asked to solve. In this essay we discuss these two reasons by investigating a particular problem that has been considered in numerous American articles during the past several years.

The problem we have chosen is rich enough to present a challenge to students at almost any level, illustrates the great variety of approaches that can be used to solve many mathematical problems, and also brings to light numerous connections among seemingly unrelated problems.

Problem: *Eight children are entered in a table-tennis tournament. If each child plays one game with each of the other children in the tournament, how many games will be played altogether?* Depending on their age, students might suggest various ways to solve this problem: acting it out, drawing eight dots to represent children and using connecting lines to show games played (see Figure 1), giving the children names and making a list (see Table 1), considering how many games there would be if there were only two children, only three children, and so on.

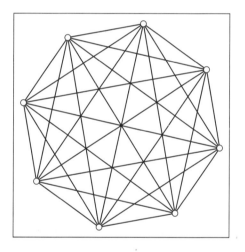

Figure 1
A diagram for solving the table-tennis problem

Table 1. *An organized list for solving the table-tennis problem*

Jenny plays:	Ann plays:	Frank plays:	and so on
Ann			
Frank	Frank		
Roger	Roger	Roger	
Lynda	Lynda	Lynda	
Jackie	Jackie	Jackie	
John	John	John	
Diana	Diana	Diana	

Or students might simply reason logically: if each of the eight children plays a game with each of the seven others, then there are 8×7 games, but because each game involves two players, there are only $56 \div 2 = 28$ different games.

---

* This discussion was prepared by Frank K. Lester and Diana V. Lambdin.

A successful teacher of problem solving encourages flexibility in students' thinking by helping them to agree on labels for the strategies listed above and then by referring to these strategies again whenever they are used in a new problem situation. Expert teachers of mathematical problem solving go even further than merely encouraging diversity in approaches and helping students to recognize and label strategies. They also encourage students to explain their solutions, to justify their choice of strategy, and to consider in retrospect what different strategies might have been employed. In addition to having children reexamine their thinking, it is often instructive to consider, as well, how the solution would change if alterations or extensions were made to the original problem.

As an example of extending a problem, let's reconsider the problem about the eight children playing table tennis. If the students have solved the problem either by acting it out or by using a diagram, they may simply have counted to get the answer. Extending the problem by asking "What if there were more people, for example, 20 or even 100?" forces a rethinking of the limitations of the initial approach and may prompt a search for a generalization that could be extended to any number of persons. It might be observed that the 20th child would play 19 other children, the 19th child would have just 18 more people to play (because he has already played the 20th child), the 18th child would play just 17 more people, and so on. This observations leads to the generalization that the sum $n + (n - 1) + (n - 2) + \cdots + 1$ will give the solution for $n + 1$ persons. But how can we find this sum? The original problem has now been transformed and extended to a new problem.

The great mathematician K. F. Gauss is credited with an ingenious solution to the problem of summing the first 100 whole numbers, a solution that can be understood—and perhaps even discovered—by young children. (Gauss himself was only a child when he discovered it.) Gauss envisioned pairing the numbers as shown in Figure 2. Because each of the 50 pairs sums to 101, the sum is 50(101) or 1/2(100)(101). In general, the sum of the first n integers can be shown to be $1/2(n)(n + 1)$.

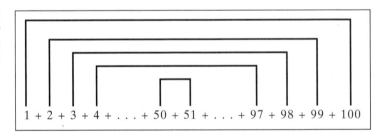

Figure 2
Gauss's method for summing consecutive whole numbers

Another interesting way to extend many problems is to attempt a geometric solution to a numerical question or vice versa. Children could visualize the sum of the whole numbers 1 to n as a set of stairs. For example, Figure 3 represents $1 + 2 + 3 + 4 + 5$. We wish to find the total number of squares that make up the staircase. If we visualize another identical set of stairs and place it upside down on top of the first, we get Figure 4. We can easily find the number of squares that make up this figure because the figure is a rectangle with dimensions 5×6. But because the rectangle was constructed of two staircases, it contains twice as many squares as we desired to count. The number of squares desired is $1/2(5 \times 6)$. More generally, for a staircase with n steps there would be $1/2(n)(n + 1)$ squares.

Teachers need to encourage children to look for connections between newly encountered problems and previously solved problems. Consider the square numbers $(1, 4, 9, 16, \ldots)$. They are called square because these are the only numbers of dots that can be arranged in square arrays (see Figure 5). It is easy to see that the nth square number is n^2. A student might ask, "What numbers would be triangular numbers?" They are those numbers of dots $(1, 3, 6, 10, \ldots)$ that can be arranged in triangular arrays (see Figure 6). The question might be asked, "How can we find the nth triangular number?" Some children might observe that we find, for example, the fifth triangular number by adding a row of 5 dots to the fourth triangular number.

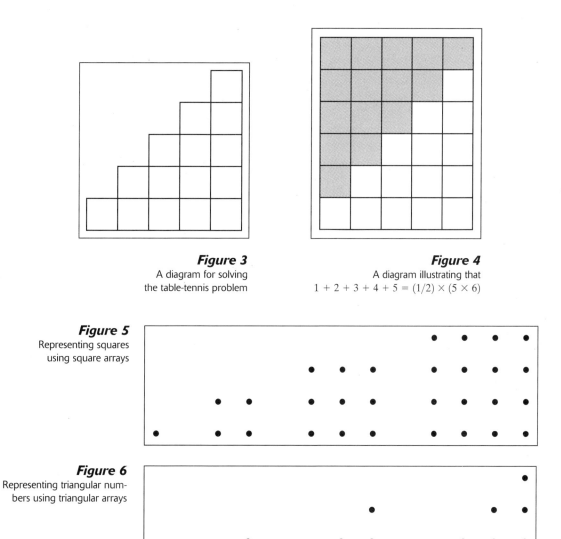

Figure 3
A diagram for solving the table-tennis problem

Figure 4
A diagram illustrating that $1 + 2 + 3 + 4 + 5 = (1/2) \times (5 \times 6)$

Figure 5
Representing squares using square arrays

Figure 6
Representing triangular numbers using triangular arrays

With this observation, it becomes clear that the nth triangular number can be found by summing the numbers $1 + 2 + 3 + \cdots n$. Other students might simply recognize the triangular arrays of dots as another way of representing the table-tennis-Gauss-staircase problem!

We have discussed the table-tennis problem and some of its many variations to illustrate several points about why students should study problem solving. This very simple problem illustrates how problems can be extended to provide students with opportunities to investigate a variety of ways to think about and to represent them. The problem variations also show how to engage children in exploring mathematical connections among problems that might otherwise seem to be completely unrelated.

Activity 1.8 The Valentine's Day Party

FYI Topics in the Student Resource Handbook
1.5 **Problem-Solving Topics**

Three married couples met at a local restaurant for a Valentine's Day dinner party. As each couple arrived, individuals meeting for the first time would shake hands. Each person met a different number of people (0–4) for the first time, except for Tom and Joyce Martin, who each met the same number of people. How many hands did Joyce Martin shake?

Activity 1.9 The Puzzle of the Hefty Hippos

Topics in the Student Resource Handbook

1.5 Problem-Solving Topics

As an incentive to lose weight, members of the Hefty Hippos Weight Watchers' Club decided that the hippo who lost the most weight by a certain date would win a prize. When the day came to determine the winner, the hippos went down to the nearby warehouse to use the heavy duty scales. But the scales started at 300 kg, more than any of them weighed (these were no ordinary hippos). What were they to do? Heloise, a hippo of some considerable mathematical prowess, came up with a solution. She said that they merely needed to weigh all possible pairs of hippos and then determine each hippo's weight from these weighings. The weights of all possible pairs were as follows (in kilograms, of course): 361, 364, 376, 377, 380, 389, 392, 393, 396, 398, 408, 411, 414, 426, 430. How much did each hippo weigh? (You may assume that all weights are whole numbers.)

Activity 1.10 Making Dice

FYI Topics in the Student Resource Handbook
1.5 Problem-Solving Topics

1. Each of the shapes given below can be folded to make a die. In each of them, three numbers are missing. Fill in the missing numbers so that the numbers on opposite faces always add up to 7.

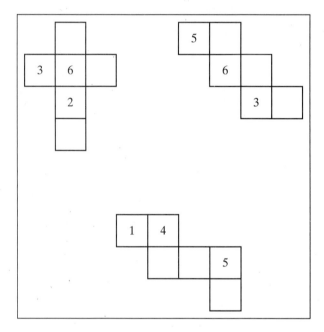

2. Create two similar shapes, perhaps with a different number of faces, and exchange them with others in your group and have them complete them.

Activity 1.10 Making Dice 27

3. Visualize a 3 × 3 × 3 cube made up of smaller 1 × 1 × 1 cubes.

   a. How many 1 × 1 × 1 cubes are there?

   b. If the large cube were painted on all the exposed faces, how many of the smaller cubes would have:

      i.   exactly four faces painted?

      ii. exactly three faces painted?

      iii. exactly two faces painted?

      iv. exactly one face painted?

      v.   no face painted?

4. What strategies did you use to answer each of the questions in (a) and (b) above?

Activity 1.11 The Tower of Hanoi

FYI Topics in the Student Resource Handbook
1.5 **Problem-Solving Topics**

Suppose you are given five disks of varying sizes. Imagine that each of these disks has a hole in the middle and that the disks are stacked on one of three pegs, with the disks stacked in decreasing size (the largest on the bottom). The goal is to transfer all five disks to one of the other pegs, using the fewest possible transfers, while observing the rules below.

1. Move only one disk at a time.

2. No disk may be placed on top of one smaller than itself at any time.

[Note: The name of this puzzle comes from a legend in which ancient Brahman priests were expected to move a stack of 64 disks according to the rules stated above. According to the legend, the world would end when the priests completed the task.]

1. How many transfers of disks are needed to move all five disks to another peg?

2. Fill in the table below, and make a conjecture about the fewest possible number of moves for N given disks. Test your conjecture for small values of N. How would you test it for large values of N? Explain your strategies for making and testing your conjectures.

Number of disks	Fewest possible number of moves

3. Suppose the legend of the Tower of Hanoi were true. If a group of Brahman priests began to move one disk per second and they began 1,000 years ago, when would the world come to an end?

Things to Know from Chapter 1

Words to Know

- devise a plan
- implement a plan
- look back and review a solution
- look for a pattern
- problem solving
- guess and check
- solve a simpler problem
- strategy
- understand a problem
- use a visual aid
- use algebra
- work backwards

Concepts to Know

- what it means to understand a problem
- what it means to devise a plan
- what it means to implement a plan
- what it means to look back and review a solution
- what it means to be systematic in solving a problem
- what it means to persevere in solving a problem
- what it means to monitor your thinking when solving a problem
- what it means to generalize a solution

Procedures to Know

- guess-and-check strategy
- solve-a-simpler-problem strategy
- use a visual-aid strategy
- use algebra strategy
- work-backward strategy
- work systematically

Exercises and More Problems

Exercises

1. 1, 2, 4, ____, ____, ____ Find the next three terms.

2. A, B, B, A, B, B, A, ____, ____, ____ Find the next three terms.

3. 5, 6, 14, 29, 51, 80, ____, ____, ____ Find the next three terms.

4. 2, 3, 9, 23, 48, 87, ____, ____, ____ Find the next three terms.

5. □, □□, □□□, □□□□, _____, _____, _____ Find the number of segments required for each of the next three terms.

6. Make up two patterns where the object is to find some missing terms.

7. Name four problem-solving strategies you have learned and implemented in this chapter.

8. List and explain the four steps of problem solving that were introduced in this chapter.

Critical Thinking

9. a. Rebekah needed exactly 6 cups of water for a recipe, but she had only two cups that could measure 5 cups and 8 cups of liquid respectively. How could she use these two to get the exact amount she needs?

   b. Could Rebekah get exactly 1 cup of water using the two cups she had?

   c. What other amounts are possible to get using the two cups?

10. June and her friend bought an 8-gallon barrel of cider that they wanted to share equally. However, the only other containers that they had were a 5-gallon container and a 3-gallon container. How could they use these to separate the cider?

11. Suppose you have an inexhaustible supply of 5-cent and 8-cent stamps.

    a. Which of the following amounts can you make up using the stamps you have? Show the combinations that you would use to make up the amounts that you can.

    7¢, 13¢, 14¢, 17¢, 18¢, 31¢, 41¢, 43¢, 52¢, 78¢, $1.64

    b. List all the amounts between 1¢ and $1 that cannot be made using these stamps.

12. Compare problems 9 and 11.

    a. What similarities do you see between the two problems?

    b. What differences do you see?

13. Moses had $23,560 in his bank account at the beginning of this year. Each month, starting with January, he withdraws half the amount of money that is left in the account.

    a. How much money will be left in his account at the end of October?

    b. When will the money in the account become less than $10? Less than $1? Show your methods of obtaining the solutions clearly. You may give all the answers correct to the nearest cent.

14. You have eight coins that look alike, but one is heavier than the others. Using a balance scale, what is the minimum number of weighings that you need to determine the heavier coin? Explain your answer.

15. Adam wants to make three waffles. He has a griddle iron that can hold at most two waffles at a time, and each side of a waffle takes three minutes to get cooked. What is the minimum amount of time that Peter will take to cook all the three waffles?

16. A spider is trying to climb a wall that is 15 feet high. In each hour, it climbs 3 feet but falls back 2 feet. In how many hours will it reach the top of the wall? Explain your answer.

17. Which of the following numbers is larger? Explain your answer.

$$(1 + 2 + \cdots + 1000) \times 999 \quad \text{or} \quad (1 + 2 + \cdots + 999) \times 1000$$

18. Sara inspected the horses and chickens in her farm one morning. Afterward, she could not remember how many animals of each kind she had, but she knew that the total number of these animals was 27, her own age in 1997. She also knew that the total number of legs of the animals was 70, the year (1970) in which she was born. How many horses does Sara have? You may assume that each animal has the usual number of legs. Outline clearly the strategy you used to answer this question.

19. Three boxes with lids are placed on a shelf. Of the three boxes, one contains all white balls, one contains all red balls, and one contains a mix of red balls and white balls. The three boxes are labeled, "White Balls," "Red Balls," and "Red Balls and White Balls," but all three are placed incorrectly. By pulling out a single ball from any one of the three boxes, how can you determine the correct label for each box?

20. Diana has a sheet of paper that is 8 1/2 inches × 11 inches in size. How can she cut off a square that is 6 inches × 6 inches, using only the paper?

21. Six friends Eric, Harry, Kevin, John, Tom, and Max are comparing their weights. Kevin is 32 pounds lighter than Harry. Tom is 26 pounds heavier than Eric. Harry is 4 pounds lighter than Tom but 17 pounds heavier than John. Max is halfway between the heaviest and the lightest person. Arrange the six friends in decreasing order of their weights. If Eric's weight is 130 pounds, find the weight of the other five people.

22. The bus system of the Longwinded Bus Company runs in a strange way. On most routes, the buses only run in one direction. Thus, to travel from one town to another, passengers may have to go through several other towns. The following statements show the bus routes between 10 destinations.

- From Stratford, buses travel to Temple and Hastings. (This means that passengers can go from Stratford to Temple or Hastings but that they cannot travel back on the same route).
- From Temple, buses go to Mapletown.
- From Blissville, buses go to Oakwood, and buses also go from Oakwood to Blissville.
- Buses go from Pinesville to Brackport and Williamsburg.
- Mapletown gets buses only from Temple.
- From Morgantown, buses travel to Stratford and Pinesville.
- Buses go from Hastings to Blissville.
- From Mapletown, buses go to Williamsburg.
- From Blissville, buses go to Pinesville.
- Buses go from Williamsburg to Stratford.
- From Stratford, buses go to Brackport.

Using the Longwinded Bus Company, how would you travel:

    a. from Stratford to Blissville?

    b. from Morgantown to Mapletown?

    c. from Williamsburg to Oakwood?

    d. from Pinesville to Oakwood?

    e. from Hastings to Williamsburg?

    f. from Mapletown to Brackport?

    g. from Stratford to Pinesville?

    h. from Brackport to Williamsburg?

    i. from Pinesville to Mapletown?

    j. from Temple to Hastings?

23. Place six plants along the four walls of a room so that each of the walls has the same number of plants placed along it.

24. Consider a billiard table that is arranged in the form of a grid as follows: A ball that starts its path from one corner of the table at an angle of 45° will traverse diagonally across each square until it hits a side. It will then reverse its direction and once again travel at an angle of 45° until it hits another side. This will continue until the ball hits a corner. An example of such a path is given below. In this case, the width of the table is 3 units, and its length is 2 units.

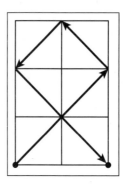

   a. Draw a set of seven tables, each table having width of two units; the lengths 1, 2, 3, 4, 5, 6, and 7 units respectively. In each case, show the complete path that the ball would travel if it started from the bottom left corner and was struck at an angle of 45° to the side.
   b. State at least three patterns that you notice in the above set of pictures.
   c. Using these pictures, where will the ball will end up for a table that has dimensions 27 × 2 units? 100 × 2 units? 1002 × 2 units? Explain how you obtained your answers.

25. A population of bacteria doubles every day. On the 30th day, the population is 20 million. On what day was it 5 million?

26. Tyler used 2,989 digits to number the pages of a book. How many pages does the book have? Justify your answer.

27. Sara, Cathy, and Tina have just finished playing three games. There was only one loser in each game. Sara lost the first game, Cathy lost the second game, and Tina lost the third game. After each game, the loser was required to double the money of the other two. After three rounds, each woman had $24. How much did each have at the start?

28. Four people want to cross a bridge on a very dark night. They all begin on the same side. You have to get all four of them across to the other side as quickly as possible. There is only one flashlight. A maximum of two persons can cross the bridge at one time. Any group that crosses, either 1 or 2 people, must have the flashlight with them. The flashlight must be walked back and forth; it cannot be thrown, and so on. No one can be carried. Each person walks at a different speed. Person A takes 1 minute to cross, person B takes 2 minutes, person C takes 5 minutes, and person D takes 10 minutes. A pair must walk together at a rate of the slower person's pace. What is the least amount of time needed to get all four people across the bridge?

29. Each of 117 crates in a supermarket contains at least 80 and at most 102 apples. What is the smallest number of crates that must contain the same number of apples?

Extending the Activity

30. Suppose you are playing *Poison* and there are 6 tiles left on the table. If it is your turn, how many tiles should you take?

31. Suppose you are beginning a game of *Poison* and there are 29 tiles on the table. Do you want to play first or second? How many tiles should you take to be sure you will win?

32. Suppose you are playing *What's My Number?* Here is a list of the guesses and information obtained for each guess. What's the number?

Guess	Correct Digits	Correct Places
123	1	0
234	1	0
345	1	1
456	0	0
567	1	1
678	1	0
789	1	0

33. Suppose you are working with the Tower of Hanoi puzzle and you have 6 disks on the left-most peg and you want to move them to the right-most peg. What is your first move?

34. Suppose you have 11 disks on the left-most peg and you want to move them to the right-most peg. What is your first move?

35. Given the number of disks and the target peg, can you predict the first move? Explain your answer.

36. Write down an eight-digit number consisting of the two 1s, two 2s, two 3s, and two 4s such that the 1s are separated by one digit, the 2s are separated by two digits, and so on.

37. Pick any single-digit number. Multiply this number by 273. Now, multiply the answer by 407. What do you see? Try this with several different single-digit numbers.

    a. Explain why you get such an answer each time.

    b. Suggest two different numbers (other than 273 and 407) that would work in the same way.

38. Visualize a $4 \times 4 \times 4$ cube.
    a. How many smaller cubes is it made up of?
    b. If the large cube were painted on all the exposed faces, how many of the smaller cubes would have
       i. exactly four faces painted?
       ii. exactly three faces painted?
       iii. exactly two faces painted?
       iv. exactly one face painted?
       v. no face painted?
    c. What strategies did you use to answer the questions in a. and b. above?

Writing/Discussing

39. Looking for relationships between problems is an important part of doing mathematics. The solution to *Stacking Cereal Boxes* could have been very helpful in solving *Laying Blocks in a Patio* or vice versa. Explain the connections you see between these two problems.

Exercises & More   Writing/Discussing   **37**

40. Discuss the strategies that you used in constructing numbers in Activity 1.6.

41. Write an account of how your group solved either the *Valentine's Day Party* problem or the *Puzzle of the Hefty Hippos* problem. Then write an efficient explanation of the problem solution.

42. Discuss what you have learned about doing mathematics so far in this course. In particular, consider what you have learned about various problem-solving strategies, making and testing conjectures, and evaluating solution efforts. Include in your discussion your thoughts about how this course has been different from previous mathematics courses.

CHAPTER TWO

Numeration

CHAPTER OVERVIEW

Among the most important topics in elementary school mathematics are those concerned with systems of recording and naming numbers: numeration systems. In this chapter, you will learn about the properties of a good numeration system by comparing and contrasting properties of several systems that have been used by various cultures throughout history. Also, to help you appreciate the issues involved when young children first begin to learn properties of our base-ten (decimal) numeration system, your group will develop one of your own. Finally, you will investigate two special properties of numeration systems—place value and base—to help you understand better the roles these properties play in the base-ten system we use.

BIG MATHEMATICAL IDEAS

Generalizing, problem-solving strategies, decomposing, mathematical structure, multiple representations

NCTM PRINCIPLES & STANDARDS LINKS

Number and Operation; Problem Solving; Reasoning; Communication; Connections; Representation

Activity **2.1** Early Numeration Systems
2.2 The Hindu-Arabic Numeration System
2.3 Comparing Numeration Systems
2.4 Mathematical & Nonmathematical Characteristics of Systems
2.5 Creating a Numeration System
2.6 Exploring Place Value Through Trading Games
2.7 Converting from One Base to Another
2.8 Place Value and Different Bases
2.9 Solving a Problem with a Different Base
2.10 Computations in Different Bases

ACTIVITY 2.1 Early Numeration Systems

FYI Topics in the Student Resource Handbook
2.1 Characteristics of Numeration Systems

Read about different numeration systems in the *Student Resource Manual*. Then, complete the following activity that explores some early numeration systems.

1. The use of tally marks to record numbers was a common numeration system used by some early communities. This system is as follows:

Tally System	Our System
/	1
//	2
///	3
////	4
/////	5
//////	6
///////	7
////////	8
.	.
.	.
.	.

   How would you write 13 with the tally system? 24? 56? 104?

2. The early Chinese-Japanese Numeration System was a base-ten system. The system was as follows:

Chinese-Japanese Numerals	Our System
一	1
二	2
三	3
四	4
五	5
六	6
七	7
八	8
九	9
十	10
百	100
千	1000

Numbers were represented by numerals written in vertical columns. For example,

Chinese-Japanese Numerals

五 (5)	一 (1)	五 (5)
十 (10)	百 (100)	千 (1000)
≈ (3)	≈ (2)	六 (6)
	十 (10)	百 (100)
	六 (6)	≈ (2)
		十 (10)
		五 (5)

| Our System | 53 | 5(10) + 3 | 126 | 1(100) + 2(10) + 6 | 5,625 | 5(1000) + 6(100) + 2(10) + 5 |

How would you write 345 with this system? 1,039? 12,678?

ACTIVITY 2.2 The Hindu-Arabic Numeration System

FYI Topics in the Student Resource Handbook

2.1 **Characteristics of Numeration Systems**
2.2 **Base-Ten Introduction**
2.3 **Place Value and Zero**

The numeration system that we use is called the Hindu-Arabic Numeration System. It has this name because both Hindus and Arabs contributed to this system. The Hindus developed an alphabet and used some letters to represent some of their digits in their numeration system. In about A.D. 600, this system developed a place-value notation, and eventually the system evolved to the system we use today. The Arabs' contribution to our numeration system lies primarily with their transmitting the information about it to other parts of the world.

The Hindu-Arabic Numeration System has six basic characteristics. These are defined below, followed by questions for you to answer.

Definition 1: A numeration system has a **base** if it reflects a process of repeated grouping by some number greater than one. This number is called the base of the system, and all numbers are written in terms of powers of the base.

What base does the Hindu-Arabic Numeration System use? Justify your answer.

Definition 2: A numeration system is a **place value** system if the value of each digit is determined its position in the numeral.

Give an example to illustrate why the Hindu-Arabic Numeration System is a place-value system.

Activity 2.2 **The Hindu-Arabic Numeration System**

Definition 3: A numeration system is **multiplicative** if each symbol in a numeral represents a different multiple of the face value of that symbol.

Give an example to illustrate why the Hindu-Arabic Numeration System is multiplicative.

Definition 4: A numeration system is **additive** if the value of the set of symbols representing a number is the sum of the values of the individual symbols.

Give an example to illustrate why the Hindu-Arabic Numeration System is additive.

Definition 5: A numeration system has a **zero** if there is a symbol to represent the number of elements in the empty set.

Explain the use of the zero in the numeral 906.

Chapter 2 *Numeration*

Definition 6: A numeration system is a **unique representation** system if each numeral refers to one and only one number.

What would be the disadvantages of a system in which some numerals did not represent unique numbers?

Discuss as a group the relationships among a number, a numeral, and the name of a number.

ACTIVITY 2.3 Comparing Numeration Systems

FYI Topics in the Student Resource Handbook

2.1 Characteristics of Numeration Systems
2.2 Base-Ten Introduction
2.3 Place Value and Zero

1. What are some advantages and disadvantages of the tally numeration system?

2. Explain how the Egyptian Numeration System is an additive system.

3. What advantages does the Egyptian System have over the tally system?

4. Explain how the Babylonian Numeration System used the idea of place value. What does each place represent in this system?

5. Why does the absence of a symbol for zero make the Babylonian Numeration System sometimes hard to use?

6. Explain how the Chinese-Japanese Numeration System is a base-ten system.

7. What are some advantages and disadvantages of the Chinese-Japanese Numeration System?

8. The Roman Numeration System was an additive system with subtractive and multiplicative features. Explain how these two features are illustrated in this system.

9. Does the Roman Numeration System have a place-value scheme? Explain your answer.

ACTIVITY 2.4 Mathematical and Nonmathematical Characteristics of Systems

Topics in the Student Resource Handbook

2.1 **Characteristics of Numeration Systems**
2.2 **Base-Ten Introduction**
2.3 **Place Value and Zero**

1. How many separate digits does the tally system have? The Egyptian? The Babylonian? The Chinese-Japanese? The Roman? The Hindu-Arabic?

System	Number of digits
Tally	
Egyptian	
Babylonian	
Chinese-Japanese	
Roman	
Hindu-Arabic	

2. Fill in the chart below for each of the numeration systems we have studied, noting whether or not the systems have the listed characteristics.

	Tally	Egyptian	Babylonian	Chinese-Japanese	Roman	Hindu-Arabic
Mathematical Characteristics						
Additive						
Multiplicative						
Has base						
Place value						
Symbol for zero						
Nonmathematical Characteristics						
Unique representation						
Convenient, easy to use						
Economical in terms of number of distinct symbols						
Ease with which it can be learned						

ACTIVITY 2.5 Creating a Numeration System

FYI Topics in the Student Resource Handbook
2.1 Characteristics of Numeration Systems
2.2 Base-Ten Introduction
2.3 Place Value and Zero

In your group, create your own numeration system that is different from the Hindu-Arabic system. Include several of the following aspects:

1. Provide a table of symbols basic to your system and the Hindu-Arabic equivalent of each symbol.

2. Have examples of larger numbers represented by the numerals of your system so that your method of recording is clear.

3. Give a rationale that supports the design of your system, and indicate the characteristics your system has.

4. Indicate which of the basic operations (addition, subtraction, multiplication, division) are possible in your system, and give examples showing how such operations would be performed.

ACTIVITY 2.6 Exploring Place Value Through Trading Games

FYI Topics in the Student Resource Handbook
2.3 Place Value and Zero

This activity involves playing two games called—310 (Fifty-Two) and Give Away—using base four blocks. In preparation for playing the games, you need to become familiar with names for the blocks on the table. The smallest block is called a unit. The next largest block is called a long, the next is a flat, and the largest block is called a cube.

310 (Fifty-Two)
[Read the explanation of the first game, and answer questions 1 and 2 before playing the game. Answer 3 and 4 after you have played the game.]

Each person starts with no blocks. Take turns throwing the two dice and finding the sum or difference of the numbers on the dice. The sum or the difference is the number of unit pieces you must add to or remove from the amount you have. You must trade smaller pieces for equivalent larger pieces to obtain the fewest number of pieces. The first player to reach the equivalent of 52 unit pieces, without going over, is the winner. Record your totals after each transaction in the chart below. As you play the game, try to think about what mathematical concepts are present in this game.

# of cubes	# of flats	# of longs	# of units	total # in units

1. What is the fewest number of throws of the two dice you would need to win a flat? What would they be?

2. What is the largest number of throws of two dice you would need to win a flat? Why?

3. What is the fewest number of unit pieces that are needed to play this game if you were going to package and sell the game for four players? Why? (Note: An adequate supply of cubes, flats, and longs will also be provided.)

4. What mathematical concepts are illustrated in this game?

Give Away

[Read the explanation of this game and answer questions 1–3 before playing. Answer questions 4 and 5 after playing the game.]

Each player starts with two flats. Take turns throwing the dice and finding the sum or the difference of the numbers on the two dice. The sum or difference is the number of unit pieces you must add or return. The winner is the first player to return all the blocks by the exact throw of the dice. Record your totals after each transaction in the chart below.

# of cubes	# of flats	# of longs	# of units	total # in units

1. What is the fewest number of throws of the two dice you would need to give away all of your blocks? What would they be?

Activity 2.6 *Exploring Place Value Through Trading Games* **51**

2. What is the largest number of throws of the two dice you would need to give away all of your blocks? Why?

3. Answer questions 1 and 2 for this game when it takes seven units to make a long.

4. Using the same blocks that you are using, a student had recorded 132 in her table. If she traded in all of her blocks for units, how many would she have?

5. What mathematical concepts are illustrated in this game?

ACTIVITY 2.7 Converting from One Base to Another

FYI Topics in the Student Resource Handbook

2.2 Base-Ten Introduction
2.4 Decimal System and More on Base Ten

1. Change these base-ten numerals into numerals in the given base.

Base 10	Base 2	Base 5	Base 6
6			
21			
45			
39			

2. Change the numerals below to base-ten numerals.

Base N	Base 10
11_{two}	
35_{six}	
104_{five}	
17_{nine}	

Activity 2.7 *Converting from One Base to Another*

3. Find the largest number from each group. Be able to justify your answer.

   a. 5_{six} 11_{four} 101_{three}

   b. 122_{three} 112_{four} 76_{eight}

   c. ET_{twelve} 101_{eight} $11,110_{two}$

   d. 325_{twelve} 523_{six} $10,122_{three}$

ACTIVITY 2.8 Place Value and Different Bases

 Topics in the Student Resource Handbook

2.2 Base-Ten Introduction
2.3 Place Value and Zero
2.4 Decimal System and More on Base Ten

1. The Hindu-Arabic Numeration System is both a base-ten and place-value system. What does this mean?

2. How does a numeration system with base two and place value differ from the Hindu-Arabic System?

3. How would you change a base-ten numeral to a base-two numeral? Hint: Doing an example may help you develop a strategy.

4. How would you change a base-two numeral to a base-ten numeral?

ACTIVITY 2.9 Solving a Problem with a Different Base

FYI Topics in the Student Resource Handbook
2.2 Base-Ten Introduction
2.3 Place Value and Zero

The ACME Potato Chip Problem

The ACME Potato Chip Company received six freight cars supposedly full of potatoes in 100-lb bags. It was learned that the automatic weighing machine was broken for a while and that some of the freight cars were full of 90-lb bags of potatoes. Ms. Jones said, "Let's load some sacks of potatoes from each freight car into our truck. Then, in one weighing, we will locate the freight cars containing sacks with the 90-lb bags. Let's take 1 bag of potatoes from the first freight car, 2 from the second, 4 from the third, 8 from the fourth, 16 from the fifth, and 32 from the sixth. So, we will have a total of 63 bags. If all the sacks weigh 100 lb, the correct answer from the weighing should be 6,300 lb more than the truck. Suppose the answer is 5,870 lb more than the truck (i.e., the potatoes in the truck are 430 lb too light). Because each light sack differs from 100 lb by 10 lb, there are 43 light sacks in the truck."

How could Ms. Jones determine which freight cars had bags of potatoes weighing 90 lb instead of 100 lb? Try to generalize this procedure.

ACTIVITY 2.10 Computations in Different Bases

FYI Topics in the Student Resource Handbook

2.2 Base-Ten Introduction
2.3 Place Value and Zero

1. Do the following computations in the base that is indicated.

$$312_{four} \quad TE3_{twelve} \quad 100_{two} \quad 156_{seven}$$
$$+233_{four} \quad +99_{twelve} \quad -11_{two} \quad -64_{seven}$$

$$23_{four} \quad 45_{six} \quad 78_{nine}$$
$$\times 30_{four} \quad \times 32_{six} \quad \times 234_{nine}$$

2. Illustrate (by drawing base-ten blocks) the division of 57 by 2.

3. How would you divide 234_{five} by 12_{five}? Explain your answer in terms of flats, longs, and units.

4. Find the whole-number base indicated by the letter b.

   a. $67_{ten} = 61_b$

   b. $12_{ten} = 1100_b$

   c. $234_{ten} = 176_b$

5. Change the following base-ten numerals to the indicated base, and place your answers from left to right in the corresponding squares. If your work is correct, the answers should read the same horizontally and vertically.

   a. $486 = \underline{}_{five}$

   b. $1{,}064 = \underline{}_{six}$

   c. $848 = \underline{}_{seven}$

   d. $298 = \underline{}_{six}$

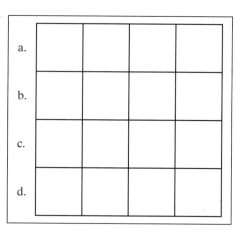

Things to Know from Chapter 2

Words to Know

- additive
- base
- expanded notation
- face value
- multiplicative
- number
- numeral
- numeration system
- place value
- subtractive
- unique representation

Concepts to Know

- what it means for a numeration system to have a base
- what it means for a numeration system to have place value
- what it means for a numeration system to be multiplicative
- what it means for a numeration system to be additive
- what it means for a numeration system to have a zero
- what it means for a numeration system to have unique representation
- the relationships among a number, a numeral, and the name of a number
- the relationship between a number written in base ten and the same number written in another base

Procedures to Know

- representing a number in various numeration systems
- representing a number in various bases
- converting a numeral from base ten to another base
- converting a numeral from one base to base ten
- performing arithmetic operations with numerals in bases other than ten

Exercises & More Problems

Exercises

1. Complete the following table.

Hindu-Arabic	Babylonian	Egyptian	Mayan	Roman
72				
	⟨⟨▽⟨▽▽▽			
			(Mayan symbol)	
				MCMLXV
121				
		99∩∩∩l		

2. Consider the following table. The number in the lower row represents the number represented by each dot in that column. Thus, in the top table, the number represented is 128. After studying the table, state the numbers represented in the other two tables. Then, make up two of your own, using the same pattern of numbers in the bottom row.

••	••••	••
49	7	1

••	•••	••
64	8	1

••		•••
169	13	1

3. Explain each of the following characteristics of our base-ten numeration system. Define what each characteristic means, and illustrate its meaning using the rational number 254.71.

   a. it has place value

   b. it is multiplicative

   c. it is additive

   d. it has unique representation

4. What is the place value of 7 in the numeral 25,746, 0 in the numeral 70,822, and 2 in the numeral 342?

5. Write a number in which 5 has the place value thousand, 3 has the place value hundred, and 0 has the place value unit.

6. a. What is the place value of 7 in 23753_{eight}?

   b. Write a number in base seven in which the digit 3 represents 3 times 49.

   c. Write the following numbers, using expanded notation: 356475, 1101001_{two}, $5T390_{twelve}$ (the letter T represents the number 10 in base ten), 21322_{five}

7. Convert the following numbers from the base they are written in to base ten.
   a. 7542_{nine}
   b. 21122_{three}
   c. 563_{eight}
   d. $E9323_{twelve}$ (E is used as a digit that represents the number 11 in base ten)

8. Convert the following numbers from base ten to the base indicated.
   a. 689 to base three
   b. 4955 to base eleven
   c. 201 to base five

9. Find x.
   a. $127_{nine} = 411_x$
   b. $321_{four} = 111_x$

10. Which of the following could be used to represent the base ten number 24?
    a. XXIV
    b. 44_{five}
    c. 20_{twelve}
    d. 1011_{two}
    e. all of the above
    f. a, b, and c

11. Use cubes, flats, longs, and units to explain how to perform the following subtraction:
    $20_{four} - 13_{four}$

12. Explain how to do the following subtraction using cubes, flats, longs, and/or units:
    $1220_{four} - 33_{four}$

13. How many cubes, flats, longs, and/or units in base 3 would you need to represent the number 35 in base 10? How many would you need in base 6?

14. Use the least number of quarters, nickels, and pennies necessary to make 87¢.

15. How would you write 324.132_{five} in expanded notation?

16. How would you write $(abcd.efg)_{\text{base } n}$ in expanded notation?

17. Numerals in Braille are written using a combination of dots, in a cell that is two dots (across) by three dots (down). All numerals are preceded by a backward L dot symbol. The following table shows the basic symbols used in Braille.

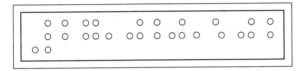

The following are examples of some numbers written in Braille numerals.

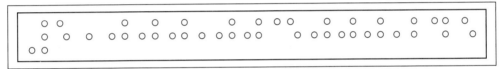

27,020,502

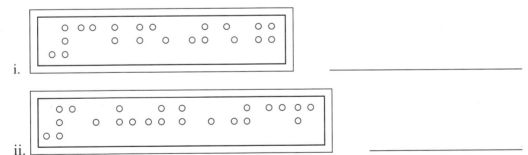

5,000,003,000,245

a. Write down the following Braille numerals in the Hindu-Arabic System.

i.

ii.

b. Write down the following Hindu-Arabic numbers in Braille numerals

i. 60,203,378

ii. 345,200

18. Complete the following addition table. The numerals are written in base four.

+			21
20		33	
	100		
	122		132

19. Complete the following addition table. The numerals are written in base six.

+			43
21		102	
	105		
	122		134

20. Write $9 \cdot 12^7 + 11 \cdot 12^4 + 10$ as a base-twelve numeral.

21. Write $7 \cdot 11^5 + 6 \cdot 11^2 + 3$ as a base-eleven numeral.

22. Perform the following operations as indicated:

   a. $8475T_{eleven} + 94T48_{eleven}$

   b. $1{,}101{,}001_{two} + 1{,}001{,}001_{two}$

   c. $7{,}437_{eight} - 5{,}476_{eight}$

   d. $TE78_{twelve} - 9{,}365_{twelve}$

   e. $3{,}323_{four} \cdot 23_{four}$

   f. $645_{seven} \cdot 36_{seven}$

For #23–24, find the sum, and state the base being used. Some helpful conversions are as follows:

3 tsp = 1 tbsp; 16 tbsp = 1 cup; 2 cups = 1 pint; 2 pints = 1 quart; 4 quarts = 1 gallon;

12 in. = 1 ft; 3 ft = 1 yd; 1,000 mm = 1 m; 1,000 m = 1 km.

23.
```
  1 gal   3 qt   1 cup
  1 gal          3 cups
+         2 qt   1 cup
```

24.
```
  9 hr   40 min   26 sec
  3 hr   50 min   17 sec
+        47 min   33 sec
```

For #25–26, find the difference, and state the base being used.

25.
```
  5 yd   2 ft   3 in
- 2 yd   2 ft   7 in
```

26.
```
  5 km   3 m   2 mm
- 2 km   4 m   9 mm
```

27. Explain the method you used to add and subtract in #23–26 above.

Convert the following measurements.

28. 512 in. = ____ yd ____ ft ____ in.

29. 39.5 ft = ____ yd ____ ft ____ in.

30. 47 tsp = ____ gal ____ qt ____ cups ____ tbsp ____ tsp

31. 79 tbsp = ____ gal ____ qt ____ cups ____ tbsp ____ tsp

32. Explain the method for each conversion in #28–31.

33. What do #28–31 have to do with doing computations in other bases?

34. Convert 5,144 seconds into the nearest number of hours, minutes, and seconds.

Critical Thinking

35. A sultan arranged his wives in order of increasing seniority and presented each with a gold ring. Next, every third wife, starting with the second, was given a second ring. Of these wives with two rings, every third one starting with the second received a third ring, and so on in this manner. His most senior and most cherished wife was the only one to receive 10 rings. How many wives did the sultan have? Try to generalize your solution.

36. Is it possible for a numeration system to be multiplicative but not have a base? Justify your answer.

37. Three boxes with lids are placed on a shelf. Three dimes and three quarters are placed in the three boxes so that there are two coins inside each box. The boxes are labeled 50¢, 35¢, and 20¢, but none of the boxes is labeled correctly. What is the minimum number of coins that you need to pull out, and from which box or boxes, if you intend to label all three boxes correctly?

38. a. Jack discovered, in climbing his beanstalk, that the giant had a numeration system based on "fee, fie, foe, fum." When the giant counted his golden eggs, Jack heard him count "fee, fie, foe, fum, fot, feefot, fiefot, foefot, fumfot, fotfot, feefotfot,...." Jack believes that the giant has 20 eggs. What are the other nine numerals that the giant used to finish the counting?

   b. If possible, characterize even and odd numbers by looking at the units digit of a given number when it is expressed in base two. In base three. In base four. In base five. In the "fee, fie, foe, fum" system.

   c. Explain your rule for determining the even numbers in base two.

39. What is the largest six-digit base-three number? Justify that this is the largest.

40. a. If all the letters of our alphabet were used as our single-digit numerals, what would be the base of this system?

   b. If *a* represented zero, *b* represented one, and so on, what would be the base-ten numeral for the "alphabet" numeral *zz?* Justify your answer.

41. A time machine carried an adventurer far into the future. Although welcomed by the natives of Septuland, he began to question their honesty when they sent him nine wives rather than the 12 that had been promised him, and only 19 free movie passes when they had promised him 25. If the Septuland natives were honest, what would explain the differences in interpreting numerals?

42. If the adventurer was promised the following number of articles by the natives, determine the number (from the base-ten adventurer's point of view) that he would actually receive.

   a. 15 suits

   b. 12 milkshakes

   c. 113 CDs

   d. 225 bananas

   e. 24 rugs

   f. 66 neckties

43. What is the smallest number of whole-number gram weights needed to weigh any whole-number amount from 1 to 12 on a balance scale? What about from 1 to 37 grams? What is the most you can weigh using six weights in this way?

44. A national track association makes shot puts that weight 1, 2, 3, 4, ... , or 15 pounds. They keep them in storage houses all over the United States. You have been hired to travel around the country and to make sure that the weights stamped on the shot puts are correct. Each shot put has a whole-number weight. You must check, for example, that no 4-pound shot put has been mistakenly stamped "3 pounds." What is the minimum number of weights you must carry to check each weight of shot put? What are these weights? Justify your answers.

Extending the Activity

45. Which number is larger, 5 or 8? Which numeral is larger?

46. a. Some children have trouble with reversals; that is, they confuse 13 and 31, 27 and 72, 59 and 95. What characteristics of the Hindu-Arabic Numeration System might contribute to the children's confusion?

    b. What numerals might give ancient Roman children difficulties? Why? How about Egyptian children? Why?

47. A woman had 40 sacks of gold dust that ranged in weight from 1 to 40 ounces. (No two sacks weighed the same, and there we no fractional weights.) One day, on returning from a long trip, she suspected that a thief had pilfered some of the gold in whole number of ounces, so she hires you to weigh each sack. She provides you with a balance scale and a box of 40 weights, ranging from 1 ounce to 40 ounces.

    a. If you are allowed to place weights on only one side of the balance and only sacks on the other side, what is the minimum number of weights you'll take from the box to weigh the sacks of gold and decide which sacks (1–40) had some stolen (if any)? How many ounces does each of these weights weigh?

    b. If you are given the same task but are allowed to use both sides of the balance for weights or sacks, what is the minimum number of weights you'd need to use? How heavy is each weight?

    c. Using both sides for weights but with sacks on only one side, what is the minimum number of weights you'd need to use? How heavy is each weight?

    d. With the weights in (a) above, every counting number 1–40 can be represented. To what base are these weights related? What about the weights in c.?

    e. How does this problem relate to the ACME potato chip problem?

Writing/Discussing

48. Discuss what you consider to be the three most important features of a numeration system and why.

49. Discuss what you have learned about the Hindu-Arabic numeration system after working with base blocks.

50. Explain, using units, longs, flats, and cubes, how to change a number from base six to base three without first converting it into base ten.

51. Describe a general procedure for changing a number represented in base n to base ten. Write another procedure for changing a base-ten number to base n.

52. Write an algorithm for adding two numbers that are written in a base other than ten. Note: Do not change the numbers into base ten to add them.

CHAPTER THREE

Operations with Natural Numbers, Whole Numbers & Integers

CHAPTER OVERVIEW

A solid understanding of addition, subtraction, multiplication, and division is crucial to being able to do mathematics, and these operations play central parts in the elementary school mathematics curriculum. In this chapter, you will study the various models (representations) for these operations involving natural numbers, whole numbers, and integers and explore different computational techniques (called algorithms). Special attention will be placed on the development of a deep understanding of each of the four operations and the wide range of algorithms that have been developed for doing large-number computations.

BIG MATHEMATICAL IDEAS

Mathematical structure, verifying, generalizing, using algorithms

NCTM PRINCIPLES & STANDARDS LINKS

Number and Operation; Problem Solving; Reasoning; Communication; Connections; Representation

Activity **3.1** Exploring Sets of Numbers
3.2 Sets and Their Properties
3.3 Classifying Word Problems by Operation
3.4 Integer Addition and Subtraction
3.5 Integer Multiplication and Division
3.6 Scratch Addition Algorithm
3.7 Lattice Multiplication Algorithm
3.8 Cashiers' Algorithm
3.9 Austrian Subtraction Algorithm
3.10 Russian Peasant Algorithm
3.11 Using Algorithms to Solve Problems
3.12 Operation Applications

ACTIVITY 3.1 Exploring Sets of Numbers

FYI Topics in the Student Resource Handbook
3.4 Addition Properties and Patterns (Whole Numbers)
3.9 Multiplication Properties and Patterns (Whole Numbers)
3.13 Introduction to Integers

We suggest that you read the explanations about the properties of sets discussed in this activity in the *Student Resource Manual* before beginning this activity.

1. Identify the property for integers that is illustrated in each example.

   a. $4 + 0 = 4$
   b. $5 \cdot 6 = 6 \cdot 5$
   c. $5 \cdot 1 = 5$
   d. $(2 \cdot 7) \cdot 5 = 2 \cdot (7 \cdot 5)$
   e. $6 + 3 = 3 + 6$
   f. $7 + {-7} = 0$
   g. $5 \cdot 6 = 30$
   h. $(3 + 4) + 9 = 3 + (4 + 9)$
   i. $9 + 7 = 16$
   j. $7(5 + 2) = 7 \cdot 5 + 7 \cdot 2$

   k. Commutative Property for addition
   l. additive inverse
   m. additive identity
   n. multiplicative identity
   o. Associative Property for multiplication
   p. Distributive Property
   q. closure for addition
   r. Commutative Property for multiplication
   s. Associative Property for addition
   t. closure for multiplication

   a. ____ b. ____ c. ____ d. ____ e. ____ f. ____ g. ____ h. ____
   i. ____ j. ____

2. The set of natural numbers can be written as {1, 2, 3, 4, ... }. How does the set of whole numbers differ from the set of natural numbers?

3. What is the relationship between these sets?

4. What is the relationship between these sets and the set of integers?

5. Draw a picture that illustrates this relationship.

6. What does it mean to say that the set of integers is closed under addition? Under multiplication?

7. Is the set of integers closed under division? Justify your reasoning.

8. Is the set of integers closed under subtraction? Justify your reasoning.

9. Is the set of whole numbers closed under addition? Under multiplication? Under subtraction? Under division? Be able to justify your answers.

10. Is the set of natural numbers closed under addition? Under multiplication? Under subtraction? Under division? Be able to justify your answers.

11. What conclusions can you make concerning the relationship between the sets of natural numbers, whole numbers, and integers and the operations under which these sets are closed?

12. A set of numbers has a Commutative Property if and only if changing the order of the numbers does not change the answer when the operation is performed on them. Give an example showing that a certain operation and a set of numbers has a Commutative Property. Give a different example showing that a different operation does not have a Commutative Property.

Activity 3.1 *Exploring Sets of Numbers* **75**

13. Think of a name for this property that might be easier to remember.

14. A set of numbers has an Associative Property if and only if changing the way three or more numbers are grouped does not change the answer when the operation is performed on them. Give an example showing that a certain operation and a set of numbers has an Associative Property. Give a different example showing that a different operation does not have an Associative Property.

15. Think of a name for this property that might be easier to remember.

16. Does the set of integers have an identity element for addition? If yes, identify it. Does the set of whole numbers? The set of natural numbers?

17. Does the set of integers have an identity element for multiplication? The set of whole numbers? The set of natural numbers?

18. Which of these sets of numbers have an additive inverse for every element of the set?

19. Which of these sets of numbers have a multiplicative inverse for every element of the set?

20. What do you think is the definition of an *additive identity?* A *multiplicative identity?*

21. What do you think is the definition of an *additive inverse?* A *multiplicative inverse?*

22. Why are the Commutative Properties for addition and multiplication so important for our use of the set of integers? What if these properties were not true for the set of integers?

23. What if the set of integers was not closed under addition, subtraction, and multiplication? Of what consequence is it that the set of integers is not closed under division?

ACTIVITY 3.2 Sets and Their Properties

FYI Topics in the Student Resource Handbook
  3.4 **Addition Properties and Patterns (Whole Numbers)**
  3.9 **Multiplication Properties and Patterns (Whole Numbers)**

1. Below is a table showing the operation & for the set 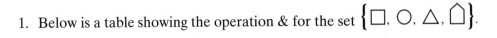.

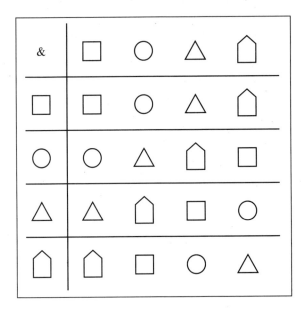

   a. Is this set closed under &? Why or why not?

   b. Does the set have a Commutative Property for &? Associative Property for &? Justify your answers.

c. Is there a & identity? If so, what element is the identity?

d. Does each element have a & inverse? If so, identify the inverses.

2. * is an operation on the set of whole numbers where $a * b$ means $2a - b$. Answer questions a, b, c, and d of item #1 for the operation * over the set of whole numbers.

3. Below are the six different ways in which two copies of an equilateral triangle with vertices 1, 2, and 3 can be placed, one covering the other. R1, R2, and R3 are clockwise rotations about the center of the triangle, while F1, F2, and F3 are flips (reflections) about a corresponding vertex respectively.

$$R1 = \begin{vmatrix} 1 & 2 & 3 \\ 1 & 2 & 3 \end{vmatrix}$$

$$R2 = \begin{vmatrix} 1 & 2 & 3 \\ 2 & 3 & 1 \end{vmatrix}$$

$$R3 = \begin{vmatrix} 1 & 2 & 3 \\ 3 & 1 & 2 \end{vmatrix}$$

$$F1 = \begin{vmatrix} 1 & 2 & 3 \\ 1 & 3 & 2 \end{vmatrix}$$

$$F2 = \begin{vmatrix} 1 & 2 & 3 \\ 3 & 2 & 1 \end{vmatrix}$$

$$F3 = \begin{vmatrix} 1 & 2 & 3 \\ 2 & 1 & 3 \end{vmatrix}$$

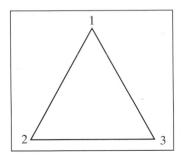

This chart shows the relationships between the rotations and the flips, using the operation that does the rotation or the flip in the column first and then the one in the row.

	R1	R2	R3	F1	F2	F3
R1	R1	R2	R3	F1	F2	F3
R2	R2	R3	R1	F2	F3	F1
R3	R3	R1	R2	F3	F1	F2
F1	F1	F3	F2	R1	R3	R2
F2	F2	F1	F3	R2	R1	R3
F3	F3	F2	F1	R3	R2	R1

What properties (closure, Commutative, Associative, identity, inverse) are valid for this set of rotations and flips?

ACTIVITY 3.3 Classifying Word Problems by Operation

Topics in the Student Resource Handbook

3.3 Addition with Whole Numbers
3.6 Subtraction with Whole Numbers
3.8 Multiplication with Whole Numbers
3.11 Whole-Number Division

Group these story-problem situations by operation. Then group them within the operation into sets of the same type of problem.

1. I have 8 marbles, and Tom has 3. How many more do I have than Tom?

2. I have 2 apples and 3 oranges. How many pieces of fruit do I have altogether?

3. I have 3 shirts of different colors and 4 shorts of different colors. How many different outfits can I wear?

4. I have 45 pieces of candy, and there are 9 children who will share the candy equally. How many pieces will each child receive?

5. I had 8 marbles. Now I have 2. How many did I lose?

6. I have 4 vases, and I want to put 3 roses in each vase. How many roses do I need?

7. I had 8 marbles. Then I lost 5. How many do I have left?

8. I have 45 pieces of candy. I am going to give 5 pieces of candy to each child in a group. How many children will receive candy?

9. A marching band has 8 rows of band members, with 7 members in each row. How many people are in the band?

10. Penguin's Ice Cream shop has 6 flavors of ice cream and 8 different toppings. How many different one dip cones are possible?

11. The library has 80 books on dogs, and there are 10 students who will check out the books, each taking the same number. How many books can each student take?

12. A company owns 7 old buses and 20 new buses. How many buses are there altogether?

13. Barb finished 14 of the homework problems, and Jim finished 9. How many more did Barb complete than Jim?

14. A gardener planted 16 rows of tulip bulbs, with 8 bulbs in each row. How many bulbs did she plant altogether?

15. Bob was taking 6 classes. Then, he dropped 2 because of time. How many classes was he still taking?

16. Joan has 5 sisters, and she is giving each of them 3 tickets to a play. How many tickets does she need?

17. Donald Trump owned 14 cars. Now he has 7. How many did he sell?

18. An airline is advertising 66 discount tickets to persons who buy pairs of tickets. How many people can buy a pair of tickets?

ACTIVITY 3.4 Integer Addition and Subtraction

FYI Topics in the Student Resource Handbook

3.13 **Introduction to Integers**
3.14 **Integer Addition**
3.15 **Integer Subtraction**

1. Think of some situations in everyday life where we use negative numbers. Write down a list of some of these situations.

Is it possible to represent these situations with positive numbers only? Are negative numbers really necessary?

2. Negative integers are often depicted in the following way on the integer-number line:

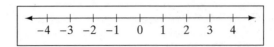

Thus, −3 is the opposite of 3, and 9 is the opposite of −9, and so on.

a. Using this number line, how would you model addition involving negative integers? For example, how would you model 3 + (−2)? (−3) + 2? (−2) + (−3)?

b. How would you model subtraction on this number line? For example, how would you represent 12 − (−7)? (−5) − 8? (−6) − (−6)?

3. For each of the following statements, write a mathematical expression that uses addition as the only operation. Then, find the answer.

   a. A certain stock registered the following gains and losses in a week: First, it rose by 7 points; then it dropped 13 points; then it gained 8 points; then it gained another 6 points; and it finally lost 8 points. What was the net change in what the stock was worth during the week?

   b. In a certain city, the temperature was 40° Fahrenheit on Sunday. It rose by 12°F on Monday, dropped by 5°F the next day, dropped another 3°F the next, and finally rose by 9°F. What was the temperature on the final day of this record?

   c. In a football series, a team gained 5 yards, lost 7 yards, gained 5 yards, and finally lost 4 yards. What was the total loss or gain?

   d. In a Las Vegas casino, a gambler gained $200, lost $115, lost another $285, and gained $245. What was his total loss or gain?

4. Jennifer and Harry have overdraft protection at their bank. At the end of a month, Jennifer's account shows a balance of −$875, while Harry's balance is −$1,575. Who has the bigger debt, and by how much?

5. Tom's bank charges a service fee of $25 each time his account is overdrawn. At the beginning of a certain month, he had $1,150 in his account. He wrote checks for $250, $348, and $628. He then deposited a check for $786, withdrew $350, wrote a check for $440, and finally deposited $500.

   a. If each check (or withdrawal or service fee) that Tom writes is written as a negative integer and each deposit as a positive integer, write a statement representing Tom's transactions.

   b. What was his final balance?

ACTIVITY 3.5 Integer Multiplication and Division

FYI Topics in the Student Resource Handbook
3.16 **Integer Multiplication**
3.17 **Integer Division**

We modeled addition and subtraction of integers using number-line models. It is harder to construct such models for either multiplication or division. For example, we may represent $2 \cdot (-3)$ in the following way on the number line.

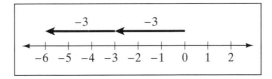

However, it is not possible to use this model to represent, say, $(-2) \cdot 3$ primarily because it does not make sense to think of this expression as 3 taken -2 times. Therefore, to be able to consider such multiplication, we will look at pattern models, such as the one given below.

$$3 \cdot 3 = 9$$
$$2 \cdot 3 = 6$$
$$1 \cdot 3 = 3$$
$$0 \cdot 3 = 0$$
$$-1 \cdot 3 = -3$$
$$-2 \cdot 3 = -6$$

Here, as numbers in the extreme left column decrease by 1 at each step, the numbers in the extreme right column decrease by 3 each time. Thus, we get $(-2) \cdot 3 = -6$. We can look at similar patterns to consider multiplication of two negative integers.

1. Construct a six-step pattern model (using the one shown above as an example) to show $(-3) \cdot (-5) = 15$.

2. Susan is recording the results of a chemical reaction every 10 minutes, starting at 1 P.M. The temperature of the reaction is being controlled so as to drop by 2° Fahrenheit every minute, and the temperature at 1 P.M. is 60° F.

   a. Write the temperatures for the first five minutes of the reaction.

   b. When will the temperature have reached 0° F?

   c. Show Susan's temperature column for the reaction until 2:30 P.M.

Division on the set of integers can be defined as the reverse operation of multiplication. Thus, $40 \div (-5) = -8$ because $(-8) \cdot (-5) = 40$. This approach may also be regarded as the missing-factor approach because essentially we are considering $? \cdot (-5) = 40$ or $(-5) \cdot ? = 40$.

3. How could you construct a pattern model for division by a certain integer? How would this pattern model be similar to the one you constructed in #1 for multiplication? How would it be different?

4. Construct an eight-step pattern model for division by −4. Begin with 16 ÷ (−4), and decrease the dividend by 4 each time.

5. Two divers record their depths under the sea as −55 ft and −44 ft.

   a. What can be regarded as the 0 level for the divers?

   b. What is the average of their depths?

6. Justin's business account shows a balance of −2400 (in dollars) after three business transactions. If he lost equal amounts in each of these transactions, what was his loss in each of them (assuming he had a balance of $0 prior to these transactions)?

ACTIVITY 3.6 Scratch Addition Algorithm

Topics in the Student Resource Handbook

3.5 Addition Algorithms (Whole Numbers)

Your instructor will lead you in doing a computation using an algorithm for addition called "scratch addition." This algorithm allows one to do complicated additions by doing a series of additions that involve only two single digits.

$$\begin{array}{r} 4\ 2\ 3_{\text{five}} \\ 3\ 2\ 0_{\text{five}} \\ +1\ 4\ 4_{\text{five}} \\ \hline \end{array}$$

Find the following sum using the scratch algorithm.

$$\begin{array}{r} 2\ 5\ 6_{\text{seven}} \\ 4\ 4\ 0_{\text{seven}} \\ +\ \ \ 2\ 3_{\text{seven}} \\ \hline \end{array}$$

Why does the scratch algorithm work?

ACTIVITY 3.7 Lattice Multiplication Algorithm

FYI Topics in the Student Resource Handbook
3.10 **Multiplication Algorithms**

One algorithm for multiplication is called lattice multiplication. Your instructor will lead you in doing the computation for 14 · 23 using this algorithm.

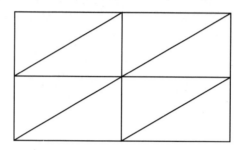

Find the product of 345 · 56 using lattice multiplication.

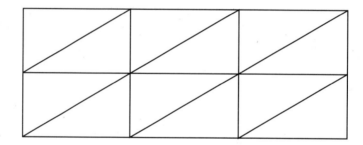

Why does lattice multiplication work?

ACTIVITY 3.8 Cashiers' Algorithm

FYI Topics in the Student Resource Handbook

3.5 **Addition Algorithms (Whole Numbers)**
3.7 **Subtraction Algorithms (Whole Numbers)**

The following is an example of the cashiers' algorithm. Suppose you buy $23 of school supplies and give the cashier a $50 bill. While handing you the change, the cashier, using the cashiers' algorithm, would say "23, 24, 25, 30, 40, 50." How much change did you receive?

Now you are the cashier. The customer owes you $62 and gives you a $100 bill. What will you say to the customer, and how much money will you give back?

Why does the cashiers' algorithm work?

ACTIVITY 3.9 Austrian Subtraction Algorithm

FYI Topics in the Student Resource Handbook

3.7 Subtraction Algorithms (Whole Numbers)

The Austrian subtraction algorithm consists of the following steps:

$$\begin{array}{rrr} 5 & 2 & 7 \\ -4 & 9 & 8 \\ \hline \end{array} \rightarrow \begin{array}{rrr} 5 & 2 & 17 \\ -4 & 10 & 8 \\ \hline \end{array} \quad \text{(add 10 to top and bottom)}$$

$$\begin{array}{rrr} 5 & 12 & 17 \\ -5 & 10 & 8 \\ \hline & 2 & 9 \end{array} \quad \text{(add 100 to top and bottom)}$$

Use the Austrian subtraction algorithm to find $721 - 348$.

Why does the Austrian subtraction algorithm work?

ACTIVITY 3.10 Russian Peasant Algorithm

FYI Topics in the Student Resource Handbook

3.10 **Multiplication Algorithms**

One algorithm for multiplication is the Russian peasant algorithm. This algorithm is illustrated below through the computation of $27 \cdot 51$.

Halving		Doubling
27	·	51
13	·	102
~~6~~	·	~~204~~
3	·	408
1	·	816

Notice that the numbers in the first column are halved (disregarding any remainder) and that the numbers in the second column are doubled. When 1 is reached in the halving column, the process is stopped. Next, each row with an even number in the halving column is ignored (crossed out), and the remaining numbers in the doubling column are added. Thus, $27 \cdot 51 = 51 + 102 + 408 + 816 = 1377$.

Use the Russian peasant algorithm to find $74 \cdot 18$.

Why does the Russian peasant algorithm work?

ACTIVITY 3.11 Using Algorithms to Solve Problems

FYI Topics in the Student Resource Handbook
- 3.5 Addition Algorithms (Whole Numbers)
- 3.7 Subtraction Algorithms (Whole Numbers)
- 3.10 Multiplication Algorithms (Whole Numbers)
- 3.12 Division Algorithms

1. The sums below are base-six representations, and in each case the addends have three digits. Recreate each problem and its solution. Under the given conditions, is there more than one possible arrangement of digits?

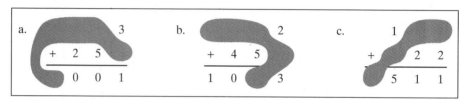

a.
```
        3
 +  2   5
 _____
    0 0 1
```

b.
```
        2
 +  4   5
 _____
    1 0 3
```

c.
```
    1
 +      2 2
 _____
    5 1 1
```

2. Find the missing digits for this base-ten numeral.

```
          _ 6 2
      3 9 4 _
   + _ 8 _ 7
   _____
    _ 3 3 2 1
```

3. What strategies did you use to find the missing digits?

4. What are the values of S, M, and O? Each letter represents a unique digit throughout the problem.

```
    S E N D
+   M O R E
-----------
  M O N E Y
```

5. Are these the only values for S, M, and O? Why or why not?

ACTIVITY 3.12 Operation Applications

FYI Topics in the Student Resource Handbook

1.5 **Problem-Solving Topics**
3.5 **Addition Algorithms (Whole Numbers)**
3.7 **Subtraction Algorithms (Whole Numbers)**
3.10 **Multiplication Algorithms**
3.12 **Division Algorithms**

1. Insert the arithmetic symbols $+$, $-$, $\times$, and $\div$ between the 6s in each line to make the eight different equations true. In each case, the arithmetic operations should be performed in order from left to right, without regard to the order of operation rules.

$$5 = 6 \quad 6 \quad 6 \quad 6$$
$$8 = 6 \quad 6 \quad 6 \quad 6$$
$$13 = 6 \quad 6 \quad 6 \quad 6$$
$$42 = 6 \quad 6 \quad 6 \quad 6$$
$$48 = 6 \quad 6 \quad 6 \quad 6$$
$$66 = 6 \quad 6 \quad 6 \quad 6$$
$$108 = 6 \quad 6 \quad 6 \quad 6$$
$$180 = 6 \quad 6 \quad 6 \quad 6$$

2. Eighteen students were seated in a circle. They were evenly spaced and numbered in order. Which student was directly opposite (a) student #1? (b) student #5? (c) student #18?

3. Another group of students was seated in the same way as above. Student #5 was directly opposite student #26. How many students were in the group?

4. A large group of students is standing in a circle evenly spaced. The seventh student is directly opposite the 791st student. How many students are there altogether?

5. Choose any number. Multiply by 2. Add 5. Multiply by 5. Subtract 25. Divide by 10. Why did you get the answer that you did? Will this always work?

6. Choose any number. Multiply by 3. Add 8. Add your original number. Divide by 4. Subtract your original number. Why did you get the answer that you did? Will this always work?

7. Choose any number. Add the number that is one larger than your original number. Add 11. Divide by 2. Subtract your original number. Why did you get the answer that you did? Will this always work?

8. Make up an algorithm so that the final result will always be 7, no matter what number is chosen.

Things to Know from Chapter 3

Words to Know

- addition
- algorithm
- associativity
- closure
- commutativity
- distributivity
- division—repeated subtraction, sharing
- identity
- integers
- inverse
- multiplication—repeated addition, cross-product, array
- natural numbers
- negative
- operation
- positive
- subtraction—comparison, take away, missing addend
- whole numbers

Concepts to Know

- what it means for a set of numbers to be closed under an operation
- what it means for a set of numbers to have commutativity under an operation
- what it means for a set of numbers to have associativity under an operation
- what it means for a set of numbers to have an identity under an operation
- what it means for each element of a set of numbers to have an inverse under an operation
- what it means for a set of numbers to have one operation to distribute over another
- relationships among the sets of natural numbers, whole numbers, and integers
- why particular algorithms work

Procedures to Know

- determining what properties hold for a set of numbers
- classifying word problems by operation type
- performing arithmetic operations with integers
- determining why particular algorithms work

Exercises & More Problems

Exercises

1. Use numbers to illustrate the following properties of integers:

    a. Associative Property of addition
    b. Commutative Property of multiplication
    c. identity property for addition
    d. multiplicative inverse property

2. Define operation @ on the set of whole numbers by $x @ y = 3x - 2y$.

    a. Is @ commutative? Explain.

b. Is @ associative? Explain.
c. Is the set of whole numbers closed under @? Explain.
d. Does the set of whole numbers have an identity element under @? Explain.

3. Given below is an addition chart for the set {a, b, c}.

+	a	b	c
a	a	b	c
b	b	c	a
c	c	a	b

a. Does this set have an additive identity? Explain.
b. Is addition commutative on this set? Explain.
c. Does b have an inverse? Explain.

4. Given the table below, answer the following:

@	a	b	c
a	b	c	a
b	a	b	c
c	c	a	b

a. Is the set {a, b} closed under @? Explain.
b. Is the set {a, b, c} closed under @? Explain.
c. Does the operation @ have an identity element? Explain.
d. Is @ commutative? Explain.
e. Does c have an inverse? Explain.

5. Give an example that illustrates whether or not division on the set of whole numbers is distributive over subtraction.

6. Place parentheses, as needed, to make the following statements true.

a. $3 + 9 - 4 + 2 = 6$
b. $3 + 9 - 4 + 2 = 10$
c. $7 \cdot 2 + 5 = 49$
d. $7 + 6 - 4 \cdot 2 = 5$
e. $3 \cdot 5 + 2 = 21$

7. Create a word-problem situation illustrating the following:

a. addition
b. missing-addend subtraction
c. cross-product multiplication
d. sharing division
e. array multiplication

8. Create a word-problem situation illustrating the following:
   a. comparison subtraction
   b. take-away subtraction
   c. repeated-addition multiplication
   d. repeated-subtraction division

9. For each of these word-problem situations, identify the operation type (a–i). The operation types may be used more than once.

   a. addition b. comparison subtraction c. take-away subtraction
   d. missing-addend subtraction e. repeated-addition multiplication
   f. cross-product multiplication g. array multiplication
   h. repeated-subtraction division i. sharing division

   ____ I have 8 oranges and Jane has 5. How many more do I have than Jane?

   ____ I have 3 sweaters of different colors and 5 pairs of pants of different colors. How many different outfits can I wear from these clothes?

   ____ I have 72 pencils. I am going to give 3 pencils to each student in my class. How many students are in the class?

   ____ Rebekah got 10 problems correct on the quiz, and Jill got 7 correct. How many more did Rebekah get correct than Jill?

   ____ Moses drove 75 miles in the morning and 120 miles in the afternoon. How many miles did Moses drive?

10. Which of the following are integers? If they are, identify them as positive, negative, or neither.
    a. 13   b. −4   c. 0.5   d. 0   e. −1/4   f. 2   g. 5.75   h. 1

11. Illustrate the following sums:
    a. $4 + (-2)$ b. $-7 + 3$ c. $-6 + (-5)$

12. Find the following sums using scratch addition.

    a. 6 3 4
        1 5 8
          4 2
        3 9 2
      +5 1 0

    b. 6 4 5
          3 4
        3 7 9
          4 8
      +2 5 2

13. Find the following products using lattice multiplication.
    a. $543 \cdot 61$ b. $1{,}802 \cdot 435$

Chapter 3 *Operations with Natural Numbers, Whole Numbers & Integers*

14. Suppose you are a cashier and a customer gives you $30 when making the following purchase. Describe how you count out the money you return to the customer.

    a. $22   b. $21.50   c. $26.80

15. Use the Austrian subtraction algorithm to find:

    a. $631 - 287$ b. $2{,}423 - 1{,}657$

16. Use the Russian peasant algorithm to find:

    a. $13 \cdot 59$ b. $23 \cdot 62$

Exercises #17–22 are written with missing numbers. The geometric shapes used to represent these numbers are defined in the statement of the problem. Use the geometric shapes to describe the solution process for each problem.

17. Sam is making a long-distance telephone call that costs △ for the first minute and ▭ for each additional minute. If Sam talks for ◯ additional minutes, what is the cost of his call?

18. The student council operates a school store that sells pencils and notebooks. The council receives a profit of △ cents on each pencil and ▭ cents on each notebook they sell. During one week, the store sold ◯ pencils and ◇ notebooks. How much profit did the council make that week?

19. Michelle is planning a pizza party. △ people will be at the party. Michelle guesses that each person will eat ▭ pieces of pizza. If each pizza is to be cut into ◯ pieces, how many pizzas should Michelle order?

20. A gym class has ◇ students. △ teams are formed with ◯ players on each team. How many students have yet to be assigned to a team?

21. A carton of ◇ bottles of soda is priced at ▭ dollars, including deposit. If Jill bought a single bottle at this rate, how much change should she receive from a ◯ dollar bill?

22. A punch is made from ◯ ounces of fruit juice and ◇ bottles of ginger ale. Each bottle of ginger ale contains ▭ ounces. How many △-ounce servings of punch will this recipe produce?

23. Express each of the numbers 1 through 10 using four 4s and any of the four basic operations.

24. Create three algebraic equations involving the four basic arithmetic operations, each of whose solution is a negative integer.

25. Create two word problems representing real situations that involve operations with integers.

Critical Thinking

26. a. If a set is closed under addition, must it also be closed under multiplication? Explain your answer or give a counterexample.
    b. If a set is closed under multiplication, must it also be closed under addition? Explain your answer or give a counterexample.

27. A given set contains the number 1. What other numbers must also be in the set if it is closed under addition? Explain.

28. An eighth-grade student claims she can prove that subtraction of integers is commutative. She points out that if a and b are integers, then $a - b = a + -b$. Because addition is commutative, so is subtraction. What is your response?

29. A student claims that if $x \neq 0$, the $|x| = -x$ is never true because absolute value is always positive. What is your response?

30. True or false;
    a. Every natural number is a whole number.
    b. Every natural number is an integer.
    c. Every whole number is a natural number.
    d. Every whole number is an integer.
    e. Every integer is a whole number.
    f. The set of additive inverses of the whole numbers is equal to the set of integers.
    g. The set of additive inverses of the integers is equal to the set of integers.

31. Find the following sums using scratch addition. The numerals are written in base six.

    a. 2 3 4
        1 5 3
          4 2
        3 0 2
       +5 1 0

    b. 1 4 5
          3 4
        3 2 1
          4 0
       +1 5 2

32. Find the following products using lattice multiplication. The numerals are written in base five.
    a. 342 · 41
    b. 1,203 · 230

33. Complete the pyramid below so that each number represents the sum of the two numbers directly below it.

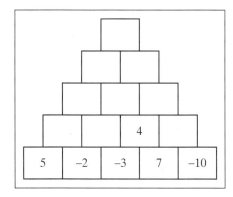

34. Complete the pyramid below so that each number represents the product of the two numbers directly below it.

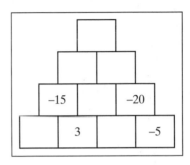

35. An elevator in an office building is at the fourth floor. It goes up 10 floors from there, comes down 6 floors, goes up 15 floors, comes down 3 floors, and goes up another 2 floors, from where it is 4 floors below the top of the building. How many floors does the building have?

36. Sara, Tyler, Diana, Joanna, and Adam were standing for election for the posts of the president, the vice president, and the treasurer. Assuming that each of them was eligible for all the three posts and that one and only one person could hold each of the posts, in how many ways can the three posts be filled? Describe the strategies you used to solve the problem.

Extending the Activity

37. a. Is the set of whole numbers closed under addition if 7 is removed from the set? Explain your answer.
    b. Is the set of whole numbers closed under multiplication if 7 is removed from the set? Explain your answer.
    c. Answer the same questions if the number to be removed is 6 instead of 7. Explain your answer.

38. Identify the property or properties illustrated in the statements below.
    a. $17 + 34 = 34 + 17$
    b. 7 times 5 is a whole number
    c. $25 \cdot 34 = 34 \cdot 25$
    d. $7 + 0 = 7$
    e. $3 \cdot 1/3 = 1$
    f. $(5 \cdot 3) \cdot 23 = (3 \cdot 5) \cdot 23 = 3 \cdot (5 \cdot 23)$
    g. $7 + 0 = 0 + 7$
    h. $5 + (-5) = 0$

39. Below is an illustration of another algorithm that can be used to find the product of two whole numbers.

    $54 \cdot 13 = 54 \cdot (10 + 3)$

    $ = 54 \cdot 10 + 54 \cdot 3$

    $ = (50 + 4) \cdot 10 + (50 + 4) \cdot 3$

    $ = 50 \cdot 10 + 4 \cdot 10 + 50 \cdot 3 + 4 \cdot 3$

    $ = 500 + 40 + 150 + 12$

    $ = 702$

    a. What property of numbers is used in the above algorithm?
    b. How is this algorithm similar to the standard algorithm for multiplication?
    c. Pick two three-digit numbers, and find their product using this algorithm.
    d. What aspects of this algorithm make it easy to use?

40. Explain why the standard algorithm for division works. For example,

    ```
           2 9
       12 ) 3 4 8
           -2 4
            1 0 8
           -1 0 8
                0
    ```

41. A third-grade teacher prepared her students for division this way:

    $20 \div 4$

    ```
       20
     -  4
       16
     -  4
       12
     -  4
        8
     -  4
        4
     -  4
        0         Thus, 20 ÷ 4 = 5
    ```

    How would her students find $42 \div 6$? Why does this algorithm work?

42. Make up your own algorithms for each of the four operations.

Writing/Discussing

43. Discuss why you think we use the standard algorithms that we do for addition, subtraction, multiplication, and division.

44. Pick one algorithm for an operation that you think has advantages over other algorithms for the same operation, and discuss what the advantages are.

45. Generalization is a key idea in these activities. Discuss what generalization is and why you think it is considered a key idea.

46. Write an explanation of the most efficient way to solve a problem like #4 in Activity 3.11.

CHAPTER FOUR

Number Theory

CHAPTER OVERVIEW

Number theory is a branch of mathematics that involves the study of numbers, in particular, the natural numbers. In this chapter, you will be introduced to prime and composite numbers and investigate the concepts of divisibility, greatest common divisor, and least common multiple—concepts that are fundamental to understanding operations on fractions. An especially interesting aspect of this chapter is that you will see that there are some quite challenging and valuable problems involving what appears at first glance to be a very simple branch of mathematics. A final important feature of this chapter is that you will begin to learn how to construct mathematical proofs.

BIG MATHEMATICAL IDEAS

Conjecturing, decomposing, verifying, problem-solving strategies, multiple representations

NCTM PRINCIPLES & STANDARDS LINKS

Number and Operation; Problem Solving; Reasoning; Communication; Connections; Representation

Activity **4.1** The Locker Problem
4.2 Searching for Patterns of Factors
4.3 Factor Feat: A Game of Factors
4.4 Classifying Numbers According to Prime Factorization
4.5 E-Primes
4.6 Twin Primes and Prime Triples
4.7 Divisibility Tests
4.8 Divisibility in Different Bases
4.9 Factors and Multiples
4.10 A Different Way of Counting
4.11 Operations with Modular Arithmetic
4.12 Mystery Numeration System
4.13 Figurate Numbers
4.14 Number Ideas: Proofs Without Words
4.15 The Fibonacci Sequence
4.16 Pascal's Triangle

ACTIVITY 4.1 The Locker Problem

FYI Topics in the Student Resource Handbook

4.1 **Prime and Composite Numbers**
4.2 **Prime Factorization and Tree Diagrams**

Students at an elementary school decided to try an experiment. When recess is over, each student will walk into the school one at a time. The first student will open all of the first 100 locker doors. The second student will close all of the locker doors with even numbers. The third student will change all the locker doors with numbers that are multiples of three. (Change means closing locker doors that are open and opening lockers that are closed.) The fourth student will change the position of all locker doors numbered with multiples of four; the fifth student will change the position of the lockers that are multiples of five, and so on. After 100 students have entered the school, which locker doors will be open?

ACTIVITY 4.2 Searching for Patterns of Factors

FYI Topics in the Student Resource Handbook

4.1 Prime and Composite Numbers
4.2 Prime Factorization and Tree Diagrams

Complete this grid to show the factors of each number from 1 through 50. What patterns do you see in the grid?

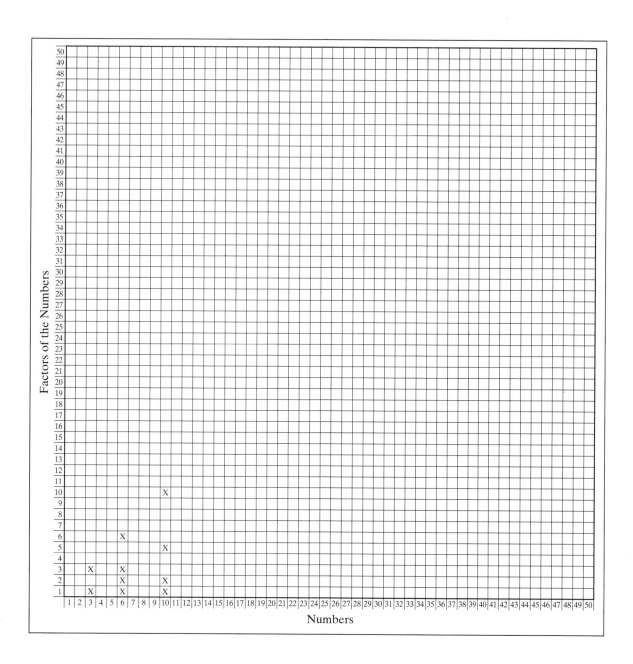

112 Chapter 4 Number Theory

Numbers that have exactly two factors are known as prime numbers. Except for the number 1, numbers that are not prime are called composite numbers and can always be broken down into a product that consists entirely of prime numbers. This product is known as the prime factorization of the number. For example, the prime factorization of 6 is 2×3, and the prime factorization of 60 is $2 \times 2 \times 3 \times 5$.

Complete the table below.

Number	List of all factors	# of factors	prime or prime factorization
2	1, 2	2	prime
3	1, 2	2	prime
4	1, 2, 4	3	2×2
5			
6			
7			
8			
9			
10			
11			
12			
13			
14			
15			
16			
17			
18			
19			
20			
21			
22			
23			
24			
25			

ACTIVITY 4.3   Factor Feat: A Game of Factors

FYI Topics in the Student Resource Handbook
- **4.1 Prime and Composite Numbers**
- **4.2 Prime Factorization and Tree Diagrams**

								2	3	4	5	6	7	8	9	10
11	12	13	14	15	16	17	18	19	20
21	22	23	24	25	26	27	28	29	30
31	32	33	34	35	36	37	38	39	40
41	42	43	44	45	46	47	48	49	50
51	52	53	54	55	56	57	58	59	60
61	62	63	64	65	66	67	68	69	70
71	72	73	74	75	76	77	78	79	80
81	82	83	84	85	86	87	88	89	90
91	92	93	94	95	96	97	98	99	100

Rules: Play with a partner. Players alternate taking turns picking numbers and marking it (one player use a circle, and the other use an *x*). When a player has marked a number, that player also marks all the factors of that number that have not already been marked. The winner is the player with the **largest sum of marked numbers** when the game is over (when either time is called or all the numbers have been marked).

ACTIVITY 4.4 — Classifying Numbers According to Prime Factorization

FYI Topics in the Student Resource Handbook
4.1 Prime and Composite Numbers

1. Examine the numbers with exactly three and five factors. Name three numbers greater than 50 with exactly three factors.

2. Name two numbers greater than 50 with exactly five factors.

3. Predict the smallest integer that has exactly seven factors.

Activity 4.4 *Classifying Numbers According to Prime Factorization* **115**

4. Write a conjecture about the numbers that have exactly three factors.

5. Here is a conjecture about some of the numbers that have exactly five factors. Conjecture: Every number with prime factorization p^4 has exactly five factors. Does this seem true? Why or why not?

6. Do you think this conjecture is true for k^4, where k is any number? Justify your answer.

7. Predict three numbers greater than 50 with exactly six factors.

8. Write a conjecture about the numbers with exactly six factors.

9. Without writing out the list of factors for 225, explain why 225 (15 × 15) has an odd number of factors but more than three factors.

10. Now write out all the factors of 225.

11. Explain why 289 (17 × 17) has exactly three factors.

Activity 4.4 *Classifying Numbers According to Prime Factorization*

12. What are all the factors of 289?

13. Use your conjectures to fill in the following chart.

Number	Prime Factorization	Odd or Even # of Factors	Exact # of Factors
529	23 × 23		
126	2 × 3 × 3 × 7		
441	3 × 3 × 7 × 7		
169	13 × 13		
11,025	3 × 3 × 5 × 5 × 7 × 7		
841	29 × 29		

14. Write a conjecture about the numbers that have exactly seven factors.

15. Write a conjecture about the numbers that have an odd number of factors.

16. Write a conjecture about numbers that have exactly two factors.

17. Write a conjecture about numbers that have exactly four factors.

18. Test your conjecture on the examples in the chart below. Note: p and q stand for prime numbers.

Number	Prime Factorization	Exactly Four Factors?
34	2 × 17	
546	2 × 3 × 91	
95	5 × 19	
	$p \times q$	
45	5 × 9	
	$p \times p \times q$	
8	2 × 2 × 2	
27	3 × 3 × 3	
	$p \times p \times p$	

19. How would you prove that there are exactly two representations of numbers with four factors?

ACTIVITY 4.5 E-Primes

Topics in the Student Resource Handbook

4.1 Prime and Composite Numbers

Let $E = \{1, 2, 4, 6, 8, \ldots\}$. In this set there are some numbers that can only be written as a product of 1 and the number itself; they cannot be written as the product of two other elements of the set. An element of E will be called **E-prime** if it can only be expressed as a product of 1 and itself. For example, 6 is E-prime because $6 = 1 \times 6$; $6 = 2 \times 3$, but 3 is not in E. An even number will be called **E-composite** if it is not E-prime. Note: 1 is not E-prime.

1. Determine the first 10 E-primes.

2. Can every E-composite number be factored into a product of E-primes? Justify your reasoning.

3. List several even numbers that have only *one* factorization into E-primes.

4. Find an even number whose E-prime factorization is not unique, that is, an even number that can be factored into products of E-primes in at least two different ways.

5. Determine a test to decide whether an even number is E-prime.

ACTIVITY 4.6 Twin Primes and Prime Triples

FYI Topics in the Student Resource Handbook
4.1 Prime and Composite Numbers

1. Consider the following triplets of numbers:

    5, 6, 7   11, 12, 13   17, 18, 19   29, 30, 31

    The first and third numbers in each triplet are called twin primes because they form a pair of consecutive prime numbers.

    a. What do these triplets have in common?

    b. Make a conjecture about twin primes in general.

    c. Verify that this conjecture is valid.

2. Prime triples are three prime numbers of the form $p - 2, p, p + 2$. There is exactly one prime triple. Find it and justify why there are no others.

ACTIVITY 4.7 Divisibility Tests

FYI Topics in the Student Resource Handbook
4.3 **Fundamental Theorem of Arithmetic**
4.4 **Divisibility**

Many of the facts, problems, and results of elementary number theory involve the idea of divisibility. In this activity you will determine methods for determining whether a given number is divisible by another number without having to actually perform the division.

Use this page to record your work about divisibility tests.

ACTIVITY 4.8 Divisibility in Different Bases

FYI Topics in the Student Resource Handbook
4.4 Divisibility

1. Determine a divisibility test for 3 in a base-twelve system. Justify your test.

2. Determine a divisibility test for 5 in a base-six system. Justify your test.

3. Determine a divisibility test for 2 in a base-five system. Justify your test.

ACTIVITY 4.9 Factors and Multiples

FYI Topics in the Student Resource Handbook

4.5 Greatest Common Divisor
4.6 Least Common Multiple

1. Find the largest common factor of both 180 and 693.

2. Find the largest common factor of 360, 336, and 1260.

3. What strategies did you use to find these factors?

4. What is the name given for the largest common factor of several numbers?

5. Write a generalization describing how to find the largest common factor for any group of numbers.

6. Find the smallest number that will be divisible by both 80 and 36.

7. Find the smallest number that will be divisible by 126, 525, and 300.

8. What strategies did you use to find these numbers?

9. What is the name given for the smallest number divisible by several numbers?

10. Write a generalization describing how to find the smallest number divisible by a group of numbers.

ACTIVITY 4.10 A Different Way of Counting

Topics in the Student Resource Handbook

4.7 Clock Arithmetic & Modular Arithmetic

Nicole and Bob had been cooped up in the house all day because of rain and were very bored. To keep them busy, their mother gave them a problem to do. They were promised a special treat if they could give her the right answer. The problem was as follows:

Nicole was born 969 days after Bob. If Bob was born on a Wednesday, on what day of the week was Nicole born?

Nicole: Okay, let me make a chart. We'll have Monday, Tuesday, Wednesday, ... Sunday. Then, we can start counting.

So, she wrote down:

M	T	W	Th	F	S	Sun

Then she counted:

M	T	W	Th	F	S	Sun
			1	2	3	4
5	6	7	8	9	...	

Bob: Wait, Nicole, I think we can do this faster!

Nicole: (after a pause) Oh, I see what you mean! Hmm, (thinks for a few minutes) I was born on a Saturday!

Try to figure out how Bob and Nicole worked this problem out so fast. Find a way to work out similar problems starting from any day of the week and for any number of days thereafter.

ACTIVITY 4.11 Operations with Modular Arithmetic

FYI Topics in the Student Resource Handbook
4.7 Clock Arithmetic & Modular Arithmetic

1. How can addition and subtraction be represented on a clock? For example, if you have a 5-hour clock and it is 2 o'clock now, what time will it be in 7 hours? What time was it 8 hours ago?

2. Write a procedure for finding the time, given that you start at time a and let b hours elapse.

3. How is arithmetic on the 5-hour clock similar to arithmetic mod 5? How is it different?

4. Fill in these tables for addition and multiplication in mod 4.

+	0	1	2	3
0				
1				
2				
3				

×	0	1	2	3
0				
1				
2				
3				

5. Use the tables to find:

   a. $2 - 3 =$ b. $3 - 1 =$

   c. $0 - 2 =$ d. $3 \div 2 =$

   e. $2 \div 3 =$ f. $0 \div 2 =$

6. Are there any restrictions on subtracting or dividing in mod arithmetic? How does 5d. relate to dividing by zero over the set of integers?

ACTIVITY 4.12 Mystery Numeration System

FYI Topics in the Student Resource Handbook
 4.1 **Prime and Composite Numbers**
 4.4 **Divisibility**

The numeration system proposed below has a number theoretic basis. Complete the table by discovering the pattern. Hint: Look at the representation of prime numbers first. Then, look at the representation of composite numbers.

Standard Number Name	Mystery System	Standard Number Name	Mystery System
one	0	eleven	10,000
two	1	twelve	12
three	10	thirteen	100,000
four	2	fourteen	1,001
five	100	fifteen	110
six	11	sixteen	
seven	1,000	seventeen	
eight	3	eighteen	
nine	20	nineteen	
ten	101	twenty	

1. What are the advantages and disadvantages of the Mystery System?

2. For small values (e.g., 1–25), is each numeral in the Mystery System associated with a unique number? Justify your answer.

3. For large values (e.g., 2^{10}), is each numeral in the Mystery System associated with a unique number? Justify your answer.

4. Is each number associated with a unique numeral in the Mystery System?

ACTIVITY 4.13 Figurate Numbers

FYI Topics in the Student Resource Handbook
4.8 Patterns

1. **Triangular numbers** are numbers that can be represented by dots in an equilateral triangle. The first four triangular numbers are illustrated below. T_n denotes the nth triangular number.

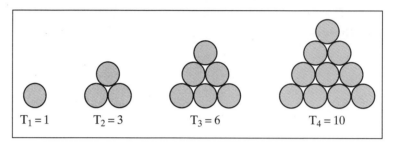

$T_1 = 1 \qquad T_2 = 3 \qquad T_3 = 6 \qquad T_4 = 10$

Find a generalization that gives the nth triangular number.

2. **Square numbers** are numbers that can be represented by dots in a square arrangement or array. The first four square numbers are illustrated below. S_n denotes the nth square number.

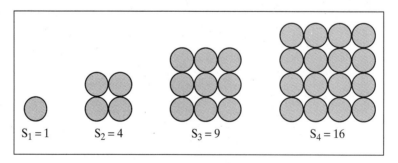

$S_1 = 1 \qquad S_2 = 4 \qquad S_3 = 9 \qquad S_4 = 16$

Find a generalization that gives the nth square number.

3. **Oblong numbers** are numbers that can be represented by dots in a rectangular array having one dimension one unit longer than the other. The first four oblong numbers are illustrated below. O_n denotes the nth oblong number.

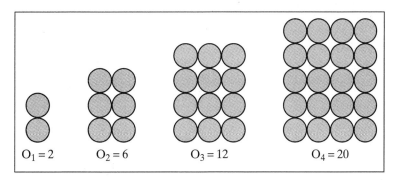

$O_1 = 2$ $O_2 = 6$ $O_3 = 12$ $O_4 = 20$

Find a generalization that gives the nth oblong number.

4. **Pentagonal numbers** are numbers that can be represented by dots in a pentagonal array. The first four pentagonal numbers are illustrated below. P_n denotes the nth pentagonal number.

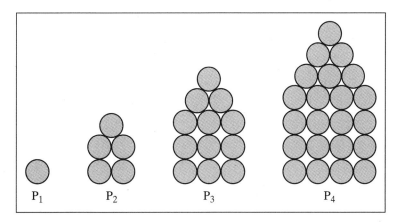

P_1 P_2 P_3 P_4

Find a generalization that gives the nth pentagonal number.

ACTIVITY 4.14 Number Ideas: Proofs Without Words

FYI Topics in the Student Resource Handbook
4.8 **Patterns**

The following illustrations depicts an example of what some mathematicians call "a proof without words."

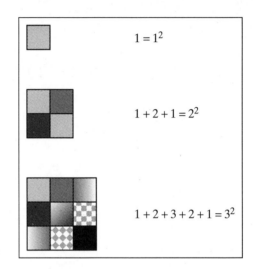

$1 = 1^2$

$1 + 2 + 1 = 2^2$

$1 + 2 + 3 + 2 + 1 = 3^2$

1. Sketch the next illustration in this pattern.

2. Generalize the result above. Then use the generalization to find the sum of the first n natural numbers.

ACTIVITY 4.15 The Fibonacci Sequence

FYI Topics in the Student Resource Handbook
4.8 Patterns

The Fibonacci Sequence is one of the most fascinating sequences of numbers that has ever been studied by mathematicians. The sequence was first suggested by Leonardo of Pisa, also called Fibonacci, in about 1202 in a book entitled Liber Abaci. In it, he proposed a problem about rabbits. The problem is as follows:

A pair of rabbits, one month old, produce a new pair every month, starting from the second month. If each new pair of rabbits does the same, and assuming that none of the rabbits ever die, calculate the number of rabbits at the beginning of each month.

1. Work this problem out for the first six months. What do you notice about the numbers?

2. Now write down the first 20 numbers in this sequence.

A sequence of numbers that follows this pattern is called the Fibonacci Sequence. One of the fascinating aspects of this sequence is that it appears in an amazing variety of creations, both natural and artificial. It appears in pine cones, sunflowers, the keys of a piano, the reproduction patterns of bees, pineapples, data sorting, and Roman poetry.

A Fibonacci sequence is also extremely interesting because of the innumerable patterns of numbers hidden in the sequence. We will consider some of these patterns in the following questions.

3. Consider the list of numbers in the Fibonacci Sequence you have written down.

   a. Which terms in the sequence are divisible by 2? Describe any patterns that you see.

   b. Which terms in the sequence are divisible by 3? Describe any patterns that you see.

   c. Which terms in the sequence are divisible by 5? Describe any patterns that you see.

   d. Which terms in the sequence are divisible by 13? Describe any patterns that you see.

e. Which terms in the sequence do you think will be divisible by 55?

4. Consider the following pattern.

   $1 + 1 = 2$

   $1 + 1 + 2 = 4$

   $1 + 1 + 2 + 3 = 7$

   $1 + 1 + 2 + 3 + 5 = 12$

   a. Describe the pattern that you notice here.

   b. Write down the next five steps in this sequence without actually performing the addition.

5. The Fibonacci Sequence can start with an arbitray first and second term. For example, we could choose 2 and 5 as the initial numbers. Then the sequence will be as follows:

$$2, 5, 7, 12, 19, 31, 50, \ldots$$

Check and see if the patterns defined above hold true in this sequence as well.

6. Study the Fibonacci Sequence, and find another pattern. Check to see if your pattern still holds when you start with two different numbers.

ACTIVITY 4.16 Pascal's Triangle

FYI Topics in the Student Resource Handbook
4.8 Patterns

Consider the pattern of numbers given below.

```
                        1
                     1     1
                  1     2     1
               1     3     3     1
            1     4     6     4     1
         1     5    10    10     5     1
      1     6    15    20    15     6     1
                        .
                        .
                        .
```

1. How do you get each successive row of this triangle?

2. Add two more rows to the triangle.

This pattern of numbers is called Pascal's triangle, after the great seventeenth-century French mathematician Blaise Pascal. The pattern was known and in use much before Pascal's time, but he is credited with making ingenious use of it in the theory of probability. The triangle may be continued indefinitely. In this activity, however, we will only look at some of the interesting number patterns hidden here, without delving into probability.

3. Find the sum of the numbers in each row. How is this sum related to that particular row?

Consider the diagonal patterns marked below.

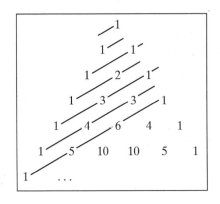

4. Find the sum of the numbers lying along each diagonal. What do you find?

5. Find and describe two other patterns in Pascal's triangle.

Things to Know from Chapter 4

Words to Know

- composite
- congruence modulo m
- divisibility
- divisibility test
- divisor
- even
- factor
- Fibonacci Sequence
- figurate number
- greatest common factor
- least common multiple
- modular arithmetic
- multiple
- odd
- Pascal's triangle
- prime
- prime factorization
- proof

Concepts to Know

- what it means for a number to be prime
- what it means for a number to be composite
- what it means for a number to be even
- what it means for a number to be odd
- what it means for a number to be divisible by another number
- what it means for a number to be a factor of another number
- what it means for a number to be a multiple of a number
- what it means for a number to be a common factor of two or more numbers
- what it means for a number to be a common multiple of two or more numbers
- what it means to count modulo m
- what it means to prove something

Procedures to Know

- finding all the factors of a number
- writing the prime factorization of a number
- classifying a number by the number of its factors
- testing whether one number is divisible by another number
- finding the greatest common factor of two or more numbers
- finding the least common multiple of two or more numbers
- performing arithmetic operations in modulo m
- representing figurate numbers symbolically
- proving an idea geometrically

Exercises & More Problems

Exercises

1. Write the prime factorization of the following:

   a. 210   b. 136   c. 54   d. 1,000   e. 2,000

2. Write the prime factorization of the following:

   a. 28   b. 72   c. 91

3. Find the prime factorization of 110 and 204. Describe the process that you used to find the prime factorization.

4. Identify the following as prime, composite, or neither:
   a. 39   b. 1   c. 59   d. 119   e. 113   f. 0   g. −3

5. Classify the following numbers as prime or composite. Explain your answer.
   a. 7   b. 16   c. 28   d. 29   e. 32   f. 43   g. 57
   h. 102   i. 116   j. 119   k. 143   l. 153   m. 171

6. Classify the following numbers as even or odd. Explain your answer.
   a. 7   b. 16   c. 28   d. 29   e. 32   f. 43   g. 56
   h. 102   i. 118   j. 125   k. 143

7. Simplify each of the following fractions. Then write a brief description of the method you used.
   a. 12/28   b. 90/105   c. $\dfrac{2 \cdot 5^3 \cdot 7 \cdot 11}{3 \cdot 5 \cdot 7^2}$

8. List five numbers that have:
   a. exactly two factors   b. exactly three factors   c. exactly four factors

9. Consider the integer 310. Does this integer have more than six factors, exactly six factors, or less than six factors? How do you know? Explain.

10. True or false:
    a. 3 is a factor of 9   b. 9 is a factor of 3   c. 9 is a multiple of 3

11. If 21 is a factor of n, what are the numbers that you can definitely say are factors of n?

12. Determine whether the following are divisible by 3. Explain your answers.
    a. 234,567   b. 19,582   c. 111,111

13. Decide whether or not the following are true using divisibility ideas. Explain your answers.
    a. 325,608 is divisible by 24   b. 1,732,800 is divisible by 40
    c. 13,075 is divisible by 45   d. 677,916 is divisible by 36

14. Pick any 10 numbers, each of which has at least four digits. Test each of them for divisibility by several of the numbers for which you've learned divisibility tests. Check your answers by actual division.

15. Find the greatest common factors for each of the following groups of numbers.
    a. 924 and 1,012   b. 864, 624, and 819
    c. 1,056 and 2,480   d. 488, 720, 968, and 704

16. Find the least common multiple of each of the group of numbers in #16.

17. Complete the following addition table in mod 4.

+	0	1	2	3
0				
1				
2				
3				

    a. What is the identity under addition in mod 4?
    b. What is the additive inverse of 3 (mod 4)?
    c. What is the additive inverse of 2 (mod 4)?

18. Complete the following multiplication table in mod 4.

×	0	1	2	3
0				
1				
2				
3				

    a. Is there a multiplicative identity? If so, what is it? Explain.
    b. Does 2 have a multiplicative inverse mod 4? If so, what is it? Explain.
    c. Does 3 have a multiplicative inverse mod 4? If so, what is it? Explain.

19. Perform the following operations modulo the number indicated in the parentheses.

    a. $3 + 9 \pmod{11}$
    b. $1 \div 3 \pmod 4$
    c. $5 \cdot 7 \pmod 8$

    d. $8 + 3 \pmod 9$
    e. $2 - 5 \pmod 8$
    f. $3 + 5 \pmod 6$

    g. $2 \div 5 \pmod 6$
    h. $5 - 7 \pmod 9$
    i. $3 \div 2 \pmod 5$

    j. $1 + 4 \pmod 5$
    k. $3 \cdot 6 \pmod 7$
    l. $4 \cdot 5 \pmod{10}$

    m. $3 - 2 \pmod 4$
    n. $1 - 5 \pmod 7$
    o. $8 \cdot 9 \pmod{11}$

    p. $4 \div 6 \pmod 8$

20. Find the additive inverses of the following. Explain your answers.

    a. 3 (mod 5)
    b. 4 (mod 6)
    c. 2 (mod 7)

21. Find the multiplicative inverses of the following. Explain your answers.

    a. 5 (mod 7)
    b. 6 (mod 9)
    c. 3 (mod 4)

22. For integers a, b, c, and k, suppose that $ac = bc \pmod k$. Show by an example that we need not have $a = b \pmod k$.

Critical Thinking

23. How many factors does p^{13} have, where p is a prime number? Explain.

24. Use the prime factorization of the given numbers to answer the following:
    a. The number 2,352 is the product of two consecutive numbers. Find them.
    b. The number 65,025 is the product of the squares of two consecutive odd numbers. Find them.
    c. The number 15,525 is the product of three consecutive odd numbers. Find them.

25. The primes 2 and 3 are consecutive integers. Is there another pair of consecutive integers both of which are prime? Explain.

26. A number is said to be *perfect* if the sum of all its proper factors is equal to the number. One such number occurs within the first 10 counting numbers. Find it.

27. Which of the following numbers is perfect?
    a. 28   b. 51   c. 256   d. 496

28. A number has 2, 3, and 5 as divisors. If it has exactly five other divisors, find the number.

29. Construct a number with exactly five divisors. Make a generalization about any number with an odd number of divisors.

30. Find the smallest positive integer with exactly six divisors. Justify why this is the smallest, and why it has exactly six factors.

31. A prime is said to be a *superprime* if all the numbers obtained by deleting digits from the right of the number are also prime. For example, the prime number 7331 is a superprime because the numbers 733, 73, and 7—obtained by removing digits on the right of 7331—are all prime.
    a. What digits cannot appear in a prime that is a superprime?
    b. Of the digits that can appear, which cannot appear as the left-most digits?
    c. Of the digits that can appear, which cannot appear as any except the left-most digits?
    d. Write down all the two-digit superprimes.

32. A trainer in an athletic department was asked to arrange the towels in the locker room in stacks of equal size. When she separated the towels into stacks of 4, one was left over. When she tried stacks of 5, one was left over. The same was true for stacks of 6. However, she was successful in arranging the towels in stacks of 7 each. What is the smallest possible number of towels in the locker room?

33. If the GCD $(x, y) = 1$, what is the GCD (x^2, y^2)? Explain.

34. For any two positive integers, a and b, it is always true that LCM (a, b) is divisible by GCD (a, b). Justify why this is true.

35. Devise a divisibility test for 12.

36. Fill in the blanks with the largest digit (0–9) that makes each statement true.
    a. 9874____ is divisible by 2
    b. 69____14 is divisible by 3
    c. 9631____ is divisible by 8

37. True or false:

    a. If a number is divisible by 2 and by 4, then it is divisible by 8.
    b. If a number is divisible by 8, then it is divisible by 2 and 4.

38. By selling cookies at 24¢ each, Jose made enough money to buy several cans of soda pop costing 45¢ each. If he had no money left over after buying the soda, what is the least number of cookies he could have sold? What number theory idea could you use to answer this question?

39. Adam was to plant evergreens in a rectangular array. He has 144 trees. Find all possible numbers of rows if each row is to have the same number of trees.

40. Rebekah was practicing addition by adding the numbers along each full week on the calendar. After a while, Rebekah saw the following pattern for finding the sum of the numbers (that represented the dates) in a week: Take the first number. Add 3. Multiply by 7. Does Rebekah's pattern work? Explain your answer.

41. Three teachers—Allison, Barbara, and Chelsea—go to the local library on a regular schedule. Allison goes every 15 days, Barbara goes every 8 days, and Chelsea goes every 25 days. If they are all at the library today, how many days from now will they all be back again? What number theory idea can be used to answer this question?

42. In the following expression, put appropriate mathematical symbols in appropriate places to make the expression true. Find as many ways as you can to do so.

    1 2 3 4 5 6 7 8 9 = 100

    (Remember, you can also leave the space blank to make a number that has more than one digit. For example, if you put nothing between 7 and 8, the number reads as 78.)

43. Here is a proof of the statement that there are an infinite number of primes. The proof can be found in many books. It is an example of a proof by indirect reasoning. In this kind of a proof, we assume a statement that is opposite in meaning to the result that we would like to prove. Based on this assumption, we logically derive a series of statements until we arrive at a contradiction. This contradiction shows that the assumption we made was not true, and hence its opposite, that is, the statement that we wanted to prove, is true.

    Statement: There is an infinite number of primes.

    Proof: Suppose there is only a finite number of primes, say, 2, 3, 5, 7, ..., p, where p is the greatest prime. Now, let us consider the number, $N = (2 \cdot 3 \cdot 5 \cdot \cdots \cdot p) + 1$. This number N is greater than 1. (Why?) Suppose N were composite. Then it would be divisible by a prime number. But none of 2, 3, 5, ..., p can divide N. (Why?) So N cannot be composite. Then N has to be prime. But this cannot be true, because N is bigger than any of 2, 3, 5, ..., p. Thus, we get a contradiction. Hence, our initial assumption that there are only a finite number of prime numbers cannot be true. Therefore, there must be an infinite number of primes.

    Supply the missing details in this proof.

144 Chapter 4 *Number Theory*

44. A student predicted (before actually checking) that the number 58 would have more than two factors and less than five factors. How do you explain this prediction?

45. A student asks you if zero is considered a factor of any number. What is your reply?

46. A student argues that there are infinitely many primes because "there is no end to numbers." How do you respond?

47. How do you see factors, divisors, multiples, primes, and composites in activities you do in your everyday life?

48. The United States Census Clock has flashing light signs to indicate gains and losses in the population. Here are the time periods of these flashes in seconds: Birth, 10; Death, 16; Immigrant, 81; Emigrant, 900. In other words, every 10 seconds there is a birth, every 16 seconds a death, and so on. Assume that these lights all started flashing at the same moment.

    a. If you saw the birth and emigrant signs light up at the same time, how many seconds would pass before they would both light together again?
    b. Suppose you saw the immigrant and emigrant signs both light at the same time. What is the shortest time before they will both be lighted together again?
    c. It is possible for the birth and death signs to light together every 80 seconds. Why?
    d. What is the increase in population during a one-hour period?

49. A recipe for a large batch of cookies calls for 5 eggs. Before baking several batches of cookies, there are a number of cartons of a dozen eggs and 3 additional eggs. After baking there is one egg left over. How many eggs were there to begin with?

50. Rapti went to the post office to buy an assortment of stamps. She told the clerk that she wanted 9-cent stamps, four times as many 13-cent stamps as 9-cent stamps, and some 3-cent stamps. She gave the clerk $5.00 and said that she wanted no change. Can her request be met?

51. A game is played with five containers into which marbles are dropped one at a time. For example, the first marble is dropped into container 1, the second into container 2, the third into container 3, the fourth into container 4, the fifth into container 5, the sixth into container 1, the seventh into container 2, and so on. If you started the game with 49 marbles, into which container would the last marble be dropped?

52. Suppose you wanted to avoid dropping the last marble into the third container. What possible number of marbles can you start with, using a range of 20 and 192? Try to generalize your answer.

53. a. Explain why the following statement is true.

    Any number in which a digit appears exactly three times (or in an exact multiple of 3) is always divisible by 3. For example, 123,123,123, or 773,133,713,313.

    b. Make a similar deduction for another single-digit number.

54. A palindrome is a number that reads the same forward and backward.

    a. Is every four-digit palindrome divisible by 11? Explain your answer.
    b. Is every five-digit palindrome divisible by 11? Explain your answer.
    c. Is every six-digit palindrome divisible by 11? Explain your answer.

d. Is every seven-digit palindrome divisible by 11? Explain your answer.
e. Make a generalization based on the above results.

55. Write a six-digit number by first writing down a three-digit number and then repeating it. For example, your number will look like 365365, or 859859. Now divide this number successively by 7, 11, and 13. Try this with several different numbers. What do you get in each case? Explain why you get this answer.

56. Find the remainder for each of the following without dividing the integers. Explain your strategies. What ideas from number theory did you use?

   a. $3{,}245 \div 5$
   b. $289 \div 3$
   c. $11^4 \div 3$
   d. $5^{32} \div 3$
   e. $2^{102} \div 5$
   f. $2^{96} \div 7$

57. In a chocolate box, the number of chocolates is such that if they are divided among seven children equally, there are two left over. If they are divided equally among five children, then there are three left over. What is the least number of chocolates that the box can contain?

58. Three pirates have a chest full of gold pieces that are to be divided equally among them. Before the division takes place, one of the pirates secretly counts the number of pieces and finds that if he forms three equal piles, then one is left over. Not being a generous man, he adds the extra piece to one pile, takes the pile and leaves. Later, the second pirate goes to the chest, divides the gold pieces into three equal piles, and finds a piece left over. He adds this piece to one of the piles, takes the pile and leaves. The third pirate then comes and does likewise. Later, the three pirates meet and divide the remaining gold pieces into three equal piles. How many gold pieces were there in the original pile?

59. When Gauss was a child, one of his mathematics teachers, in a bid to keep the children in the class busy for some time, asked them to find the sum of the first 100 natural numbers. However, Gauss came up with the answer in only a few minutes. He did so by rearranging the first 100 numbers in a way that immediately suggested the answer.

   a. Try to do the same. Describe the method you used.
   b. Use the method you described to find the sum of the first n natural numbers (where n is any natural number).

60. Moses was playing with a calculator and noticed that for

   $1 \cdot 2 \cdot 3 \cdot 4 \cdot 5 = 120,$ $2 \cdot 3 \cdot 4 \cdot 5 \cdot 6 = 720,$ $3 \cdot 4 \cdot 5 \cdot 6 \cdot 7 = 2{,}520$
   the GCD $(120, 720, 2{,}520) = 120$. Imagine Moses' excitement when he looked at
   $8 \cdot 9 \cdot 10 \cdot 11 \cdot 12,$ $9 \cdot 10 \cdot 11 \cdot 12 \cdot 13,$ $10 \cdot 11 \cdot 12 \cdot 13 \cdot 14,$ and
   $11 \cdot 12 \cdot 13 \cdot 14 \cdot 15$ and discovered that the GCD for all these seven products is 120. Was Moses just lucky, or can any numbers constructed in this way be added to the list and have the GCD for the entire list remain 120? (Hint: Recall any class discussions you may have had about factors of consecutive integers.)

Extending the Activity

61. a. Consider the integers from 26 to 50. Predict and record the integers that have the greatest number of factors.
    b. Predict and record the integers that have the least number of factors.

62. Organize the information for the integers 26–50 in a table like that in Activity 4.2.

146 Chapter 4 Number Theory

63. a. If a number is divisible by 4, is it always divisible by 12? Justify your answer.
    b. If a number is divisible by 12, is it always divisible by 4? Justify your answer.

64. a. Diana used the following procedure to find all the factors of an integer, n: first she wrote down 1 and n because their product is n; then she found the next largest factor (after 1) and its companion factor so that the product of these two is n. She continued on in the same manner. To find all the factors of this integer, what is the largest number Diana must test? Explain.
    b. Suppose Diana wanted to test whether an integer, p, was prime. What is the largest prime she must test? Explain.

65. How are mod 7 and base 7 alike? How are they different?

66. Make the addition and multiplication tables for mod 6.

67. Write three subtraction and three division problems in mod 6 and find the answers from these tables.

68. What properties [closure, commutative, associative, identity] are valid for addition mod 6? For multiplication mod 6? For addition mod n? For multiplication mod n?

69. Think of the square numbers in the following way:

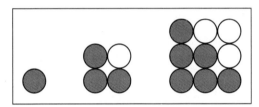

What relationship do you see between the triangular numbers and the square numbers?

70. Use the following pictures to state and prove a result about the sum of successive odd integers.

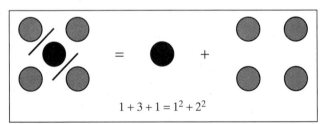

$1 + 3 + 1 = 1^2 + 2^2$

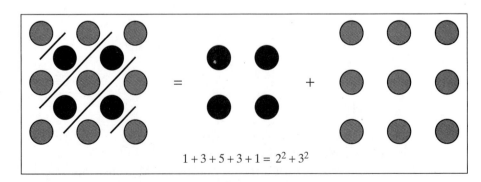

$1 + 3 + 5 + 3 + 1 = 2^2 + 3^2$

71. Consider the following series of pictures:

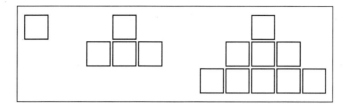

   a. Sketch the next two figures in the above pattern.
   b. What number pattern is suggested by this picture pattern?
   c. How many little squares will be there in the 10th picture in this sequence?
   d. How many little squares will be there in the 100th picture in this sequence?
   e. How many little squares will be there in the nth picture in this sequence (where n is any natural number)?

72. Consider the following series of pictures:

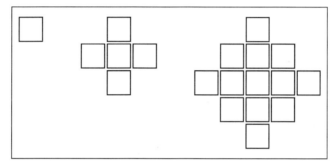

   a. Sketch the next two figures in the above pattern.
   b. What number pattern is suggested by this picture pattern?
   c. How many little squares will be there in the 10th picture in this sequence?
   d. How many little squares will be there in the 100th picture in this sequence?
   e. How many little squares will be there in the nth picture in this sequence (where n is any natural number)?

73. Generate 12 terms of the Fibonacci Sequence that begins with 1, 1, 2,.... Observe the following pattern:

$$1^2 + 1^2 = 1 \cdot 2$$

$$1^2 + 1^2 + 2^2 = 2 \cdot 3$$

$$1^2 + 1^2 + 2^2 + 3^2 = 3 \cdot 5$$

Use the terms of the sequence to predict what $1^2 + 1^2 + 2^2 + 3^2 + \cdots + 144^2$ is without performing the computation. Then use your calculator to check your prediction.

74. Extend the Pascal's triangle shown in Activity 4.16 to have 10 rows. Then predict the sums along the next three diagonals (after the ones marked in the triangle in the activity) in the triangle.

75. a. Create Pascal's triangle with seven rows. Draw a circle around one number and the six numbers immediately surrounding it. Find the sum of the encircled seven numbers.
    b. Draw several more circles as in part a. and find the sum of the encircled seven numbers.
    c. Explain how the sums of the encircled numbers are related to one of the numbers outside the circle.

76. Write out 16 terms of the Fibonacci sequence beginning 1, 1, 2, ….
    a. Look at the fourth term in the sequence. Is it odd or even?
    b. Is the sixth term odd or even? the eighth term? Look for and describe a pattern that relates the term of the sequence with evenness or oddness.
    c. Predict which of the following terms are even and which are odd: the 20th term, the 41st term, the 250th term.
    d. Look for and describe a pattern that relates the term of the sequence and divisibility by 3.

77. Prove or disprove: The sum of any 10 consecutive Fibonacci numbers is a multiple of 11.

Writing/Discussing

78. Make a concept map for factors, and explain the connections you made.

79. Write an explanation of the Locker Problem solution.

80. Make up and explain a divisibility test for 6 in base six. Generalize and explain your observations.

81. Discuss the statement that factoring and finding a product are reverse processes.

82. Explain why the Sieve of Eratosthenes works, in general, in identifying prime numbers.

83. Make a second concept map about factors. Write a reflection comparing your first and second maps, and explain your present understanding of number-theory ideas.

CHAPTER FIVE

Data & Chance

CHAPTER OVERVIEW

A great many events in the world around us involve uncertainty and chance. It is easy to find examples from business, education, law, medicine, and everyday experience. Two examples come readily to mind: (1) The weather forecaster on TV says, "There is a 70% chance of rain tomorrow"; (2) with only a very small portion of votes counted, newscasters are able to project winners of political elections and final percentages of votes with considerable accuracy. How was the 70% figure obtained? How can newscasters attain such accuracy with so little information? The branches of mathematics called probability and statistics were developed to help us deal with situations involving uncertainty and chance in a precise and objective manner. In this chapter, you will conduct to explore basic principals that underlie probability and statistics.

BIG MATHEMATICAL IDEAS

Data and chance, independence/dependence, multiple representations, mathematical modeling

NCTM PRINCIPLES & STANDARDS LINKS

Data Analysis, Statistics, and Probability; Problem Solving; Reasoning; Communication; Connections; Representation

Activity **5.1** Two Probability Experiments: Spinners & Colored Tiles
5.2 Probability Experiments with Dice & Chips
5.3 Are These Dice Games Fair?
5.4 What Would Marilyn Say?
5.5 Basic Probability Notions
5.6 Probability Models of Real-World Situations
5.7 Basic Counting Principles
5.8 Using Statistics to Summarize Data
5.9 Using Statistics in Decision Making
5.10 Looking at Variability in Data: Part I
5.11 Looking at Variability in Data: Part II
5.12 Chirping Crickets and Temperature: A Correlation Problem
5.13 Statistics and Sampling

152 Chapter 5 *Data & Chance*

ACTIVITY 5.1 *Two Probability Experiments: Spinners & Color Tiles*

FYI Topics in the Student Resource Handbook

5.1 **Probability Notions**
5.2 **Equally Likely Outcomes**

I. A Spinner Experiment

Part A. Examine the spinner given to your group. Read through the following questions. Discuss possible answers and then obtain consensus within your group about your conjectures.

1. Would you be just as likely to obtain a red as you would a blue? If not, which color is more likely to occur? Why?

2. What do you think the chances are of spinning a red in one spin?

3. What do you think the chances are of getting a blue in one spin?

Activity 5.1 *Two Probability Experiments: Spinners & Color Tiles* **153**

Test your conjectures. Record in the space below what happens as you experiment by making 25 spins. Keep in mind that you are testing your conjectures. Reflect on the results, and state any generalizations. Also discuss how experimental results may or may not change with 100 spins.

Part B. Look again at your group's spinner, and note the numbers in each sector. Read through the questions that follow, and discuss the answers to them in your group.

1. Would you be just as likely to spin a red 8 as a blue 8? Why?

2. What do you think are the chances of spinning an even number of either color? Justify your thinking.

3. What do you think are the chances of spinning a number between 3 and 8? Justify your thinking.

Test your conjectures. Record in the space below what happens as you experiment by making 25 spins. Keep in mind that you are testing your conjectures. Reflect on the results and state any generalizations. As you did in Part A, discuss how experimental results may or may not change with 100 spins.

Reflection. How did you come up with your initial answers to the questions in Parts A and B? If your initial conjectures were different from the results of your group's experimentation, why do you think they were different?

II. What's in the Bag?

DO NOT LOOK in the bag your group has been given. It contains a total of 10 objects in four different colors. You have 15 minutes to try to determine the four colors and to find the number of tiles of each color. Do this by drawing an object, recording its color, and RETURNING it to the bag. Repeat this process as many times as you wish without peeking inside the bag.

When your group has consensus, write down your prediction of the color combination. State in writing why and how you knew you had enough information to make the prediction.

Prediction:

Explanation:

Reflection. How would your predictions have changed if you had NOT RETURNED the object to the bag after drawing it out?

ACTIVITY 5.2 Probability Experiments with Dice & Chips

FYI Topics in the Student Resource Handbook
5.1 Probability Notions
5.2 Equally Likely Outcomes
5.3 Mutually Exclusive and Complementary Events
5.4 Multistep Experiments

The study of probability developed from games such as those that are played using dice. Many games are played by rolling two dice and finding the sum of numbers shown on the top face. In this activity, your group will develop and use strategies to predict some probabilities.

Part A: Getting Familiar with Dice

Roll the dice to gain some empirical data, and then determine the probabilities listed below. Be prepared to explain your reasoning.

1. List all possible sums of the two dice (these sums are called the outcomes of the rolls of the dice). List the probability of rolling each sum.

2. Find the probability of rolling a 7 or a 11 on the first roll—an automatic win in the popular game of Craps.

3. Find the probability of not rolling a 7 or a 11 on the first roll.

4. Find the probability of rolling a 1.

5. Find the probability of rolling a sum of 2, 3, or 12 on the first roll—an automatic loss in Craps.

6. Which sum that is most likely to occur? Least likely to occur?

Part B: Two Experiments with Dice and Chips

Two experiments are conducted in which chips are drawn from a bag. In both experiments there are 10 chips in the bag and each chip has a different number, from 1 to 10, written on it.

Experiment I
A person reaches into the bag and draws out a chip and records the number on it. The chips is then placed back into the bag, and the person draws out another chip and again records the number on it.

What is the probability that the sum of the two chips drawn from the bag is 12?

Experiment II
A person reaches into the bag, draws out a chip, and records the number on it. This time, the chip is *not* put back into the bag. The person then draws out another chip and again records the number on it.

What is the probability that the sum of the two chips drawn from the bag is 12?

Reflection
Think of other situations in which the probability of occurrence of an event is affected by previous events.

ACTIVITY 5.3 Are These Dice Games Fair?

Topics in the Student Resource Handbook

5.1 **Probability Notions**
5.2 **Equally Likely Outcomes**

For Game 1 and Game 2, read through the game rules; then, BEFORE YOU PLAY THE GAME, discuss with your partner, and make a prediction whether or not you think each game is fair—that is, whether both players have an equal chance of winning. Play three rounds of each game. Record your results in a table. Keep track of who wins, "odd" or "even."

Game 1: Rules

1. Choose one player to be "odd" and the other player to be "even."
2. Roll the dice, and find the absolute value of the difference of the two numbers.
3. When the difference is an odd number, the "odd" player scores one point. When the difference is an even number, the "even" player scores one point. Remember: 0 is an even number.
4. Play the game for two minutes.
5. The winner is the person with the most points at the end of two minutes.

Prediction and Reasoning:

When you have finished, discuss with your partner, and answer the following questions:

1. Do you feel more strongly about your prediction? If so, why? Do you feel less convinced about your prediction? If so, why?

2. Do you want to modify your prediction? How and why?

3. List possible strategies you could use to verify your predictions. Why would they work?

Game 2: Rules

1. Choose one player to be "odd" and the other to be "even."
2. Roll the dice, and find the absolute value of the product of the two numbers.
3. When the product is odd, the "odd" player scores one point. When the product is even, the "even" player scores one point.
4. Play the game for two minutes.
5. The winner is the person with the most points at the end of two minutes.

Prediction and Reasoning:

When you have finished, discuss with your partner, and answer the following questions:

1. Do you feel more strongly about your prediction? If so, why? Do you feel less convinced about your prediction? If so, why?

2. Do you want to modify your prediction? How and why?

3. List possible strategies you could use to verify your predictions. Why would they work?

ACTIVITY 5.4 What Would Marilyn Say?

FYI Topics in the Student Resource Handbook
- 5.1 Probability Notions
- 5.2 Equally Likely Outcomes
- 5.3 Mutually Exclusive and Complementary Events
- 5.4 Multistep Experiments
- 5.5 Counting Principles

1. A few years ago, the following problem appeared in Marilyn Vos Savant's "Ask Marilyn" column in *Parade Magazine.**

    A woman and a man (unrelated) each have two children. At least one of the woman's children is a boy, and the man's older child is a boy. Do the chances that the woman has two boys equal the chances that the man has two boys?

    Many readers wrote in to say that the chance that the man has two boys is the same as the chance that the woman has two boys. One reader insisted, "I will send $1000 to your favorite charity if you can prove me wrong."

    Determine if the man had to send $1,000 to Marilyn's favorite charity, the American Heart Association.

    [* *Parade Magazine*, October 19, 1997, page 8]

2. Here's another problem that appeared in "Ask Marilyn." The problem involves a game show and was stated as follows:

    Suppose you're on a game show, and you're given the choice of three doors: Behind one door is a car, behind the others, goats. You pick a door, say number 1, and the host, who knows what's behind the doors, opens another door, say number 3, which has a goat. He then says to you, "Do you want to pick door number 2?" Is it to your advantage to switch your choice?

    Answer A: Yes, you should switch because the first door has a 1/3 chance of winning, but the second door has a 2/3 chance of winning.

    Answer B: There is no need to switch because if the game show host opened a door and asked you if you wanted to switch, you still wouldn't know which door has the prize behind it. You now have a 50–50 chance of guessing the correct door because there are only two doors left, so it doesn't matter if you switch or not.

    Which answer do you agree with? Justify your answer.

ACTIVITY 5.5 Basic Probability Notions

FYI Topics in the Student Resource Handbook

5.1 **Probability Notions**
5.2 **Equally Likely Outcomes**
5.3 **Mutually Exclusive and Complementary Events**
5.4 **Multistep Experiments**
5.5 **Counting Principles**

> The *sample space* associated with an experiment is the set of all possible outcomes (that is, the set of all possible things that can happen) of the experiment.
>
> In an experiment, an *event* E is a subset of the sample space of the experiment. An event E is said to occur if the outcome of an experiment corresponds to one of the elements of E.

---

1. An urn is a vase whose contents are not readily visible. Your urn contains three blue marbles and two orange marbles. One marble is randomly selected, its color is noted, and it is not replaced. A second marble is then selected, and its color is noted.

   a. Draw a diagram or list the sample space, showing all possible outcomes.

b. What is the probability that

- both marbles selected are orange? Justify your answer.

- neither marble selected is orange? Justify your answer.

- at least one of the marbles selected is orange? Justify your answer.

c. How would the probabilities in b. differ, if at all, if an actual experiment was conducted and you determined the probabilities based on the data from your experiment?

2. Jamaica is a first-year college student who is only beginning to organize her laundry. Specifically, she is still throwing her socks into the drawer without pairing them. One morning when Jamaica overslept, she hurriedly grabbed two socks from the drawer without looking. Jamaica knows that in her sock drawer she has 4 white socks, 2 blue socks, 2 black socks, and 2 tan socks.

a. If Jamaica first pulls out a tan sock, what is the probability that she will select another tan sock on the next try? Explain your reasoning.

b. If Jamaica first pulls out a tan sock, what is the probability that she will *not* select another tan sock? Explain your reasoning.

c. What are her chances of pulling out either a white sock or a blue sock on the first selection? Explain your reasoning.

d. What are the chances that Jamaica pulls out a pair of matching socks if she makes two selections? Explain your reasoning.

ACTIVITY 5.6 Probability Models of Real-World Situations

FYI Topics in the Student Resource Handbook
5.1 Probability Notions
5.2 Equally Likely Outcomes
5.3 Mutually Exclusive and Complementary Events
5.4 Multistep Experiments
5.5 Counting Principles

Many real-world situations that involve chance events can be modeled using a spinner or a set of spinners. The two situations described in this activity can be modeled with a single spinner.

Situation A

Marcy plays on her school basketball team. During a recent game, she was fouled and was sent to the free-throw line to shoot a one-and-one. Use a spinner to estimate the probability that she will score 0, 1, and 2 points if she has a free throw success rate of 60%. *[Note: "One-and-one" means that the player must make the first free throw to be eligible to attempt the second.]*

1. Construct a spinner with 0.6 of the area shaded. The spinner should look something like the following:

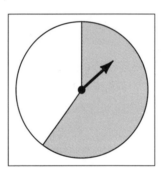

Perform an experiment consisting of 25 one-and-one trials. Be sure to follow the rules of one-and-one free-throw shooting. Record the results of your experiment in the table below.

Points	Frequency	Relative Frequency
0		
1		
2		

2. Based on the results of your experiment, how do you expect Marcy performed at the free-throw line? Why do you think this?

3. Pool the data obtained by all the groups in your class to obtain a better estimate. In what ways is this model of free-throw shooting a realistic one? What are some of its shortcomings?

Situation B

A breakfast-cereal company tries to increase its sales by putting small prizes in boxes of cereal, one prize in each box. If there are five kinds of prizes, estimate the number of boxes of cereal you would have to buy to obtain a complete set of prizes.

1. Construct an appropriate spinner to simulate this situation. Assume that the various prizes occur in equal numbers and are uniformly distributed in boxes of cereal; that is, assume that the probability that a box of cereal contains an animal of a certain type is 1/5.

2. Perform experiments that help you estimate a typical number of boxes to be opened to obtain a complete set of animals.

3. Pool the data obtained by all the groups in the class to obtain a better estimate. In what ways is this model of cereal box prizes a realistic one? What are some of its shortcomings?

ACTIVITY 5.7 Basic Counting Principles

FYI Topics in the Student Resource Handbook

5.1 **Probability Notions**
5.2 **Equally Likely Outcomes**
5.3 **Mutually Exclusive and Complementary Events**
5.4 **Multistep Experiments**
5.5 **Counting Principles**

> Whenever we want to determine the probability or chance that a certain event will occur, we need to determine how many outcomes are favorable out of the total number of possible outcomes. It is not very difficult to count the number of possible outcomes when there are relatively few; for example, if we flip three coins, it is easy to simply list all eight possible outcomes in some systematic way. Likewise, because the total number of possible outcomes is relatively small, it is not very difficult to list all possible outcomes of rolling a pair of dice. However, for many experiments this is not the case; there are simply too many outcomes to list conveniently. In this activity, you will explore ways to count all possible outcomes of an experiment. To introduce these ways to count, let's consider some situations involving the need to determine all the ways to do something.

Situation I

Suppose that in 2000 there were approximately 50,000 automobiles of a certain make sold in the United States and that each of these cars was equipped with the same type of lock. Of course, the locks on these cars were not all "keyed" alike (*Note*: Two cars are "keyed" alike if one key will work in both locks).

The primary question is: Are there enough different arrangements of the tumblers in a lock to have a different key for each of the 50,000 cars?

> **Note:** The tumblers in an ordinary lock are small metal cylinders whose height is adjusted by the depth of the notches cut into the key. If the correct key is inserted into the lock, the tops of the tumblers are aligned in such a way that the cylinder can be turned and the lock opened (or car ignition started). Of course, there also are more-complex types of locks than these.

1. Suppose keys for the cars were designed with six notch positions and six depths of notch at each position.

   a. How many different keys could be made?

   b. What if there were five notch positions and seven depths of notch at each position?

   c. What if there were seven notch positions and five depths of notch at each position?

   d. Which type of lock would be best to install in the 50,000 cars mentioned above? Explain.

2. Do we really need three-digit area codes for long distance phone calls in the United States? Could we suffice with two-digit area codes? (*Note*: Assume that the population of the United States is more than 250 million and that there are more than 200 million telephones now in use. Also, note that you may not use 0 or 1 as a first digit in the area code or in the local telephone number.)

Activity 5.7 *Basic Counting Principles* **173**

3. Suppose that a multiple-choice quiz is given with five possible responses for each item. How many different answer sheets can there be for a two-item quiz. For a five-item quiz?

4. A combination lock will open when a correct choice of three numbers is made. The numbers range from 1 through 50. How many different combinations are possible? (Is this the best name for this type of lock?)

5. Several years ago in Tulsa, Oklahoma, two service-station mechanics answered a call to start a white Ford Mustang parked on a particular city street. The owner came to the mechanics' garage, and he gave them the keys to his car and also noted that he had been unable to start the car. Unknown to the mechanics, there were two white Ford Mustangs of the same year parked in the same block and, unfortunately, not seeing two identical cars, they got into the wrong one. Because the car started up without any trouble, they decided to test drive it before reporting back to the man who had given them the keys. As they were driving off, the real owner of the car returned and called the police. The police stopped the vehicle after it had gone only a few blocks and arrested the two unsuspecting mechanics. It turned out that the key the first man had given to the mechanics worked in the locks of both cars. The police officer in charge said, "It's just one of those one-in-a-million deals that I seem to get every night or so."

"Two Identical Cars Land Mechanic in Jail." Reported in the *Daily Herald-Telephone* (Bloomington, Indiana), November 6, 1975, p. 14.

Was it a "one-in-a-million" deal? How might you go about determining how likely it would be for one key to fit the locks of two different but identical-looking cars? (*Hint*: You might want to consider special cases with a small number of cars and types of keys [e.g., four cars and two types of keys] and see if a pattern emerges.)

6. Summarize the counting method that you have used to solve problems 1–4 above.

Situation II

Anagrams are "words" that use exactly the same letters but in different orders. For example, there are two anagrams for the letters N and O: NO and ON. (Both arrangements form standard words in the English language.) There are six anagrams for the letters A, E, and R: AER, ARE, EAR, ERA, RAE, REA.

1. How many anagrams are there for the letters E, N, O, and P? (Try to figure out how many there are without actually writing out all the possible arrangements.)

2. How many anagrams are there for the letters A, E, L, P, S, and T? (PLATES, STAPLE, PETALS, PASTEL, and PLEATS are a few of the possibilities)

3. How many anagrams are possible using the letters in the word BANANA?

4. If an apple, an orange, a pear, and an avocado are placed in a row, how many arrangements of these fruits are there?

5. If the coach of a softball team wants the pitcher to bat last and the best hitter to bat in the clean-up position (i.e., fourth), how many different batting orders are possible?

6. a. The answer to questions 1 and 4 are the same. Find a third problem that has the same answer.

   b. How would you describe a method for solving problems that have the same basic features as problems 1 and 4?

Situation III

There are 10 teams in the North Central Basketball Conference (NCBC). The top three teams will automatically be invited to the national championship tournament.

1. Assuming that there are no ties, in how many different ways can the top three positions of the NCBC standings be filled at the end of the season?

2. There are 10 floats entered in a July 4th contest. The best five are to be selected and lined up for the Hooterville Independence Day parade.

    a. How many different parades of floats can there be? What does it mean for two parades to be different?

    b. If a different parade began every half-hour, 24 hours a day, how long would it be before the last parade started out? (Assume that the floats can appear in as many parades as needed.)

3. An airport bus has stopped to pick up 10 passengers to take them into the city. Four of the passengers already have their tickets; the others have to purchase theirs. If the passengers with tickets are allowed to board the bus first, in how many ways can the 10 passengers board the bus?

4. At a school cafeteria, daily lunch plates are prepared for the students consisting of one main dish, two vegetable dishes, and a dessert. The lunch plates are chosen from a menu consisting of 10 main dishes, 8 vegetables, and 13 desserts. How many different lunch plates can be prepared before the students must repeat a plate?

5. Describe a method for solving problems that have the same basic features as problems 1–4.

Situation IV

At State University, a group of seven students wishes to select a committee of four to negotiate student activity fees with the dean of students.

1. How many committees can be selected from the group of seven?

2. Five students intend to play a round-robin tennis tournament among themselves. How many matches will there be?

3. How many three-element subsets are there in a six-element set?

4. A certain state's standard license plate contains a two-digit number, followed by a letter of the alphabet, followed by a four-digit number. The leading digit of the license plate cannot be zero.

   How many different license plates can be made? Would this arrangement of letters and numerals be sufficient for your state?

5. Create a "how many possible ordered arrangements"-type problem that would have meaning for elementary school students.

6. Create a "how many possible (unordered) sets"-type problem that would have meaning for elementary school students.

7. Generalize the method for solving problems like the one you created in #6.

ACTIVITY 5.8 Using Statistics to Summarize Data

FYI Topics in the Student Resource Handbook
5.6 Statistics Notions

In this activity, some basic concepts of statistics are introduced. Examples are provided to give you an idea how these concepts can be used to summarize data.

In your *Student Resource Handbook*, there are definitions and examples of various statistical concepts. Throughout the remainder of this chapter, you will find it useful to refer to the *Student Resource Handbook* to learn about the statistical concepts and techniques that you will need to complete the activities.

Organizing & Summarizing Data

Suppose a student, Megan, earned the following grade point averages over 12 semesters of college (all have been rounded to the nearest tenth):

$$1.4, 2.0, 2.4, 3.2, 3.3, 3.3, 2.0, 3.1, 2.8, 3.0, 2.9, 3.6.$$

1. List as many methods as you can that could be used to organize these data; then display the data with each method.

2. What are the advantages and disadvantages of each method?

3. When we summarize a set of data, we look for a "typical" value (or values) that will serve as an appropriate representative of the entire set of data. The manner in which this value is determined usually depends on the situation and the use to which the value will be put. Use your *Student Resource Handbook* as a guide to help you determine the following "typical" values for the grade-point data shown above: median, mode, mean.

4. Megan's official transcript indicates that her overall grade point average is 2.716.

   a. How was this average determined, and why is it different from the values you determined in #3?

b. What additional information would you need to determine her official grade-point average?

Measuring the Spread (or Dispersion) of Data

1. Suppose Megan completed the same number of credit hours each semester. Determine the following measures of how spread out Megan's grade point averages are: range; the first, second, and third quartiles; and the standard deviation.

2. Construct a box-and-whisker plot (also called a box plot) for the grade-point-average data.

3. What does the box-and-whisker plot tell you about Megan's performance as a student over the entire 12 semesters?

Applying What You Know About "Typical"

Suppose that 25 families were selected at random and that the number of children in 1965 and 1970 were determined. The results are tabulated in the table below:

Family	1965	1970	Family	1965	1970
A	3	3	M	5	5
B	3	5	N	1	1
C	1	2	P	4	5
D	3	4	Q	6	7
E	0	3	R	3	4
F	3	3	S	1	1
G	1	3	T	2	3
H	0	0	U	1	3
I	3	4	V	4	5
J	0	0	W	0	0
K	2	2	X	4	5
L	3	4	Y	3	4
			Z	0	0

1. Consider the following statement about this table: "The typical number of children in a given family in 1965 was three, and in 1970 it was also three. However, the typical increase in family size during that period was one child."

   a. How could this statement possibly be true?

b. If the typical number of children used in the above example were the *mode*, would the same conclusion hold? The *mean*?

2. The owner of a business has an annual income of $200,000, and his 14 employees each have incomes of $25,000. How should a typical income for the 15 individuals be determined?

3. A real estate developer is planning to build a set of apartments. To save costs, he plans to construct a building in which all the apartments are exactly the same. He conducts a survey and finds that of 200 people questioned, 40 are interested in a one-bedroom apartment, 30 in a two-bedroom apartment, 20 in a three-bedroom apartment, and 110 are not interested in apartment living at all. What size apartment should he build? What sort of "average" apartment preference have you used to make your decision?

4. Many American cities have a distribution of cloudiness similar to the following.[1]

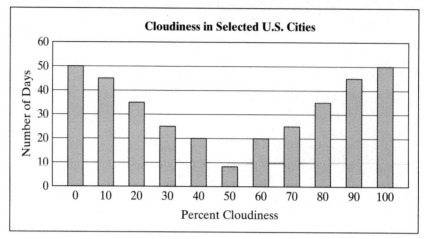

[1] S.K. Campbell (1974). *Flaws and Fallacies of Statistical Thinking.* New York: Prentice Hall, p.71.

a. How should the typical cloud cover in these American cities be described?

b. Which measure of typicality takes every data value into account and always changes when a single data value is changed? Which measures ignore extreme values?

5. Consider these test scores: 55, 85, 60, 98, 72, 80.

a. Find both the range and the mean.

b. Find a set of test scores with the same mean, but with a smaller range.

c. Find a set of test scores with the same range but with a higher mean.

6. A student took three tests of 100 points each. Her mean score was 85 and the range of her scores was 10.

   a. Find two sets of test scores that satisfy these conditions but that have different standard deviations.

   b. Find two different sets of test scores that satisfy these conditions with the same standard deviation.

   c. What do you conclude about the standard deviation as a measure of spread?

ACTIVITY 5.9 Using Statistics in Decision Making

FYI Topics in the Student Resource Handbook
5.6 Statistics Notions

Modern statistical methods play an important part in decision making in a wide variety of fields; consumer product testing, drug testing, weather forecasting, and political polling are just a few. The general approach to effective decision making that is often used involves several stages:

1. Recognize and clearly formulate a problem.
2. Collect relevant data that might help you solve the problem.
3. Organize the data in an appropriate manner.
4. Analyze and interpret the data.
5. Make decisions about the original problem.

These stages don't always take place one after another, and sometimes a stage is repeated—for example, initial analysis of the data collected may indicate a need to collect more data. The figure below illustrates the decision-making cycle.

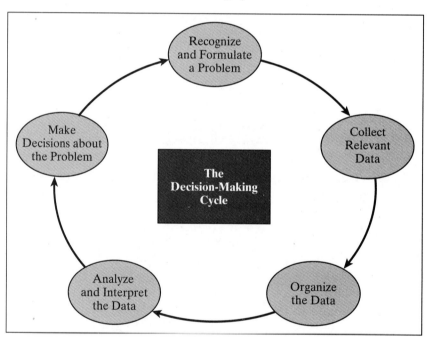

This activity will engage you in all five stages of the decision-making cycle by posing two challenges for you to consider.

Activity 5.9 Using Statistics in Decision Making

Challenge I. The Softball-Throwing Contest

Representatives of the children in the five fourth-grade classes at Elm Heights Elementary School will participate in a softball-throwing contest. One child is to be selected from each fourth-grade class.

Formulate a Problem: How should the class representative be chosen?

Suppose the class has decided that the class rep will be chosen using a three-step process: (1) Ask for volunteers to participate, (2) let each volunteer throw a softball, and (3) choose the rep on the basis of their throws.

Collect Data: What data should the class collect to solve the problem?

Suppose that there were four volunteers and that each volunteer was allowed to make five throws. The distance of each throw was measured using a trundle wheel to the nearest one-tenth meter. The data collected for the four children are shown in the following table.

Volunteers	Distance of Throws (nearest 0.1 meter)				
Robbie	27.7	23.1	22.1	23.8	26.8
Antonia	24.0	23.3	27.4	23.9	27.1
Moses	23.1	26.7	28.8	17.8	25.6
Rebekah	20.2	22.1	24.4	26.0	27.9

It is difficult to choose the class rep simply by looking at the table, so we need to organize the data in a way that will help us.

Organize the Data: How should the class organize the data to make a decision about the class representative?

1. Decide in your group how to organize these data to make it easier to select a class representative; then do the organizing by making graphs, plots, tally charts, or whatever device you think will be helpful.

Analyze and Interpret the Data: *What do the data tell us about the four children?*

2. Who is (are) the most consistent softball thrower(s)? Who is the least consistent? Should consistency be the criterion used to select the class rep? If so, why? If not, why not?

3. Which child is the "best" thrower? Which child has the best "typical" throws? (How should a typical throw be determined?)

4. Are five throws for each child an appropriate number? Should very short throws be excluded? Justify your answers.

5. a. Draw box-and-whisker plots for each child's throws and compare the plots.

   b. How do the box-and-whisker plots help you determine a typical distance? What information do these plots not give you that may be important?

   Make a Decision: Who should be the class representative?

6. Use all the information you have to make the best selection of a class representative. Support your choice as much as you can. Would you like to have different data on which to base your decision? If so, what data would be most useful to you?

Challenge II. Making Your Own Decisions Based on Data

1. Identify two other situations in which statistical thinking would be useful in making a decision that solves a problem.

   a. State the problems associated with each situation.
   b. Determine what data you would need to have to make an informed decision.
   c. Describe how you would go about collecting the data.

2. Choose one of your two problems stated above and use the decision-making cycle to solve the problem (collect data, organize the data, analyze the data, and make a decision). Prepare a report based on the decision-making process you followed.

ACTIVITY 5.10 Looking at Variability in Data: Part I

Topics in the Student Resource Handbook

5.6 Statistics Notions
5.7 Measures of Central Tendency
5.8 Data Dispersion
5.9 Representing Data

Activity. Who Is the Real Quintus Curtius Snodgrass?*

During the U.S. Civil War, the *New Orleans Daily Crescent* published a set of 10 letters containing details of the writer's military adventures as a member of the Louisiana militia. Each letter was signed "Quintus Curtius Snodgrass." Historians generally agree that the letters do refer to actual operations and activities that took place during the war, but there are no records of anyone named Quintus Curtius Snodgrass having served in the Louisiana militia. Several years after the publication of the letters, speculation began about the true identity of the writer. Some curious skeptics noted that the style of the letters was remarkably similar to that of Mark Twain, the author of such popular books as *The Adventures of Tom Sawyer* and *Adventures of Huckleberry Finn*.

> **Goal of the Activity:**
>
> To decide whether Mark Twain could have been the real author of the Quintus Curtius Snodgrass letters by looking at the variability in word-length frequencies.

Just as burglars often leave clues to their identities, authors often leave literary clues that can help identify them. For example, a particular author might tend to use approximately the same proportion of five-letter words in her or his writings, but the proportion of five-letter words that one author uses is usually very different from the proportion of five-letter words another author uses. In this activity, you will be challenged to compare the word lengths of the Q.C. Snodgrass letters with the word lengths of several of Mark Twain's writings.

Table 1 shows the distribution of word lengths for three letters known to have been written by Mark Twain. Table 2 displays the word-length distribution for the Q.C. Snodgrass letters. Use these tables to answer the questions that follow the tables.

---

* Information about the Quintus Curtius Snodgrass letters and the distribution of word lengths is taken from *Statistics in the Real World: A Book of Examples*, by R.J. Larsen & D.F. Stroup, New York: Macmillan, 1976 (out of print).

Table 1. Word-Length Distributions for Three of Mark Twain's Letters

Word Length	Letter 1 Frequency	Letter 1 Relative Frequency	Letter 2 Frequency	Letter 2 Relative Frequency	Letter 3 Frequency	Letter 3 Relative Frequency
1	74	.039	312	.051	116	.039
2	349	.185	1146	.188	496	.167
3	456	.242	1394	.288	673	.226
4	374	.198	1177	.193	565	.190
5	212	.113	661	.108	381	.128
6	127	.067	442	.072	249	.084
7	107	.057	367	.060	185	.062
8	84	.045	231	.038	125	.042
9	45	.024	181	.030	94	.032
10	27	.014	109	.018	51	.017
11	13	.007	50	.008	23	.008
12	8	.004	24	.004	8	.003
13+	9	.005	12	.002	8	.003
Total	**1,885**	**1.000**	**6,106**	**1.000**	**2,974**	**1.000**

Table 2. Word-Length Distributions for the Ten Q.C. Snodgrass Letters

Word Length	Frequency	Relative Frequency
1	424	.032
2	2685	.204
3	2752	.209
4	2302	.175
5	1431	.109
6	992	.075
7	896	.068
8	638	.048
9	465	.035
10	276	.021
11	152	.011
12	101	.008
13+	61	.005
Total	**13,175**	**1.000**

1. Use the data in Table 1 to construct a graph of the Mark Twain word-length distribution data. What would you conclude about the consistency of these data? What would you conclude about the shape of the distribution of these data?

2. Graph the data in Table 2 on the graph you have constructed above. What would you conclude about the authorship of the Snodgrass letters? Justify your conclusion.

3. Construct box-and-whisker plots of the three Mark Twain letters as well as the Snodgrass letters. What can you conclude? (Be careful! Remember that you want to compare the plots after you have constructed them.)

4. Could you resolve an authorship dispute by using only a single sample of the writing of each of two authors? Why or why not?

5. In making frequency counts such as those shown in Tables 1 and 2, what kinds of words, if any, should be excluded? Justify your answer.

6. Until DNA testing became common, in paternity suits medical information such as blood type and length of pregnancy were often used to establish whether a defendant is *not* a child's father. Why can this sort of information not be used to prove that a defendant is a child's father? Could statistical tests of authorship such as the one used in this activity be similarly "one sided"? Explain your position.

ACTIVITY 5.11 Looking at Variability in Data: Part II

Topics in the Student Resource Handbook

5.6 Statistics Notions
5.7 Measures of Central Tendency
5.8 Data Dispersion
5.9 Representing Data

Activity. Are You Liberal or Conservative? Where Do You Stand?

The political views of individuals vary tremendously across the nation and depend on such factors as ethnicity, religious affiliation, socioeconomic status, and level of education. For example, it is a widely held belief that university students have more liberal views about political issues than the population in general. Polls taken over the past several years indicate that political views also vary by geographic location. People living in a particular region of the country, as a group, may have more conservative positions on issues than people living in another region. This activity involves an investigation of this phenomenon.

The political views of students at four major state universities in four regions of the United States were determined by means of a mail survey. The survey asked questions about the students' views on four issues: (1) gun control, (2) school prayer, (3) health care, and (4) birth control. Questionnaires were mailed to a total of 15,000 students enrolled at four state universities, one each in the northeast, southeast, midwest, and far-west regions. About 6,000 students responded to the survey. On the basis of the respondents' answers, each person was classified on a five-point scale as being one of the following with respect to her or his political views:

(1) Very Liberal   (2) Liberal   (3) Middle-of-the Road
(4) Conservative   (5) Very Conservative

1. The graph below displays the political-views pattern for the 6,000 respondents from each of the four universities.

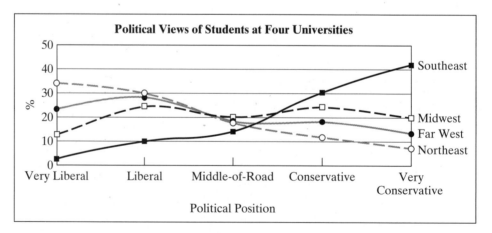

This survey was conducted by mail. Do you think this fact biases the results? If yes, in what ways might the results have been biased? If no, why do you think the results were not biased? What factors might have affected the rate of response?

2. How would you characterize the distribution of the data for respondents from the midwestern university in comparison with the other three universities?

3. Suppose we were to replace the category names—Very Liberal, Liberal, Middle-of-the-Road, Conservative, Very Conservative—with a numerical scale: 0, 1, 2, 3, 4 (0 = very liberal–4 = very conservative). Calculate the weighted mean (average) of each of the four universities. How might you use these weighted averages to characterize the respondents at each of the universities? [Refer to the *Student Resource Handbook* for information on weighted averages.]

4. What assumption is being made when the original scale (Very liberal–Very conservative) is replaced by the numbers 0–4?

5. Which type of information better characterizes the responses of the students at the four universities, the graph of the data or the numbers obtained from the weighted averages? Justify your choice.

6. What sort of response pattern do you think would be characteristic of students from southwestern state universities? Do you think the response patterns would be different for students attending private universities? Universities with religious affiliations? Justify your claims.

ACTIVITY 5.12 Chirping Crickets and Temperature: A Correlation Problem

 FYI Topics in the Student Resource Handbook
- 5.6 Statistics Notions
- 5.7 Measures of Central Tendency
- 5.8 Data Dispersion
- 5.9 Representing Data

Among all the different ways statistics is used to help us make decisions, none is more familiar to the average person than to look for how closely *correlated* (i.e., related) two sets of data are. Perhaps the most well-known example is the correlation between smoking and cancer; that is, the incidence of various kinds of cancer is higher among smokers than it is among nonsmokers. In this activity, you will learn how statistical concepts and techniques can help to establish the correlation between two collections of data.

In general, the data in a correlation problem consists of pairs of numbers (a_1, b_1), $(a_2, b_2), \ldots, (a_n, b_n)$, where a_i and b_i are two bits of information recorded for the ith person (or animal, or thing, etc.). There are two sorts of questions we want to answer about these pairs of numbers:

(a) What is the *nature* of the relationship between the *a*'s and the *b*'s?
(b) How *strong* is the relationship between the *a*'s and the *b*'s?

Let's look at an example.

Question. Is there any truth to the claim that you can determine the temperature by counting the chirps of a cricket?

Listed below are 15 pairs of data. The first number in each pair is the number of cricket chirps per second, and the second number is the measure of the temperature (in degrees Fahrenheit) for the corresponding number of chirps.

Table 1. *Cricket Chirping Frequency and Temperature**

Observation Number	Chirps per second (a_i)	Temperature, °F (b_i)
1	20.0	88.6
2	16.0	71.6
3	19.8	93.3
4	18.4	84.3
5	17.1	80.6
6	15.5	75.2
7	14.7	69.7
8	17.1	82.0
9	15.4	69.4
10	16.2	83.3
11	15.0	79.6
12	17.2	82.6
13	16.0	80.6
14	17.0	83.5
15	14.4	76.3

*From *Statistics in the Real World: A Book of Examples* by R.J. Larsen & D.F. Stroup. New York: Macmillan, 1976 (out of print).

1. Use a scatter plot to graph the data in Table 1. (Let the *x*-axis represent the chirps per second and the *y*-axis represent the temperature.) [Use the graph paper provided.]

2. Study each of the graphs below and the descriptions of the types of relationships they depict. Does the scatter plot you constructed above fit any of these descriptions? If so, which one?

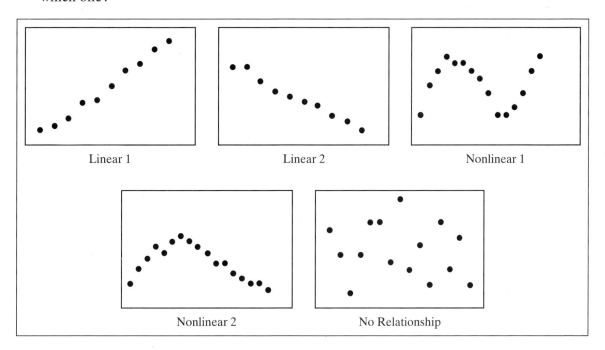

3. Based on your answer to #2, what can you say about the *nature* of the relationship between the number of cricket chirps (a_i) and temperature (b_i)? How can you justify your claim?

4. If we assume a linear 1 or linear 2 relationship between frequency of chirps and temperature, a straight line (called a line of best fit) can approximate the overall pattern in the pairs of points in your scatter plot. Also, this line can be represented by a linear equation of the form $y = mx + c$. Let's determine the equation of the line of best fit and then draw the line on your scatter plot. [Note: m and c are the slope and y-intercept of the line, respectively.]

   a. Determine each of the following: (Use your calculator!)

   $$\sum_{i=1}^{15} a_i =$$

   $$\sum_{i=1}^{15} a_i^2 =$$

   $$\sum_{i=1}^{15} b_i =$$

   $$\sum_{i=1}^{15} a_i b_i =$$

b. Use your calculator to determine $m = \dfrac{15\left(\sum_{i=1}^{15} a_i b_i\right) - \left(\sum_{i=1}^{15} a_i\right)\left(\sum_{i=1}^{15} b_i\right)}{15\left(\sum_{i=1}^{15} a_i^2\right) - \left(\sum_{i=1}^{15} a_i^2\right)^2}$ and

$c = \dfrac{\sum_{i=1}^{15} b_i - m \sum_{i=1}^{15} a_i}{15}$ (Be careful!)

c. Use the values you have just found to write the equation for the line of best fit.

d. Use the equation to draw the line of best fit on your scatter plot.

5. a. Suppose one night you counted the number of chirps the crickets were making. If the number of chirps in one minute was 1,000, approximately what was the temperature? (How might you determine a cricket's chirp rate?)

b. Write a statement that would tell someone how to determine the temperature if you know how fast a cricket is chirping.

6. The line of best fit gives only an approximation of the temperature for a given number of chirps per second.

a. Why is this line only an approximation?

b. What obvious drawback does a cricket have as a thermometer?

ACTIVITY 5.13 Statistics and Sampling

Topics in the Student Resource Handbook

5.6 Statistics Notions
5.7 Measures of Central Tendency
5.8 Data Dispersion
5.9 Representing Data

Sampling Example 1

On Feb. 18, 1993, just after President Clinton took office, a television station in Sacramento, California, asked viewers to respond to the question: "Do you support the president's economic plan?" The next day, the results of a properly conducted study asking the same question were published in the newspaper.

	Television Poll	Newspaper Survey
Yes	42%	75%
No	58%	18%
Not Sure	0%	7%

Sampling Example 2

In 1975 [advice columnist] Ann Landers asked "If you had it to do over again, would you have children?" Nearly 10,000 parents responded, with nearly 70% saying they would not. Many accompanied their responses by heart-rending tales of the cruelties inflicted on them by their children.... A nationwide random sample commissioned in reaction to the attention paid to Ann Landers's results found that 91% of the parents would have children.

[Both examples are reprinted from Pierce, D., Wright, E., & Roland, L., *Mathematics for Life*, Upper Saddle River, NJ: Prentice Hall, 1997, page 222.]

1. To what can the discrepancies between the results of the television poll and the survey in the first example be attributed?

2. Why do you think Ann Landers's question yielded such different results from those of the random sample?

Market Research

1. Market research is sometimes based on persons in samples chosen from telephone directories and contacted by telephone.

   a. What groups of people do you think will be underrepresented by such a sampling procedure?

b. How could the sample be changed to include persons and households that will be underrepresented?

c. Using the sample frame of households chosen from telephone directories, discuss how general (or not) the results might be.

d. List other ways that a sample may be chosen that will leave some groups of people underrepresented. Explain your reasoning.

2. A university employs 2,000 male and 500 female faculty members. The equal employment opportunity officer polls a random sample of 200 male and 200 female faculty members.

   a. What is the chance that a particular female faculty member will be polled? Explain.

   b. What is the chance that a particular male faculty member will be polled? Explain.

   c. Explain why this is a probability sample.

d. Each member of the sample is asked, "In your opinion, are female faculty members in general paid less than males with similar positions and qualifications?"

180 of the 200 females say "Yes."
60 of the 200 males say "Yes."

So 240 of the sample of 400 said "Yes," and the report states that, "based on a sample, we conclude that 60% of the total faculty feel that female members are underpaid relative to males." Is this a valid conclusion? Explain why or why not.

3. The method of collecting data can influence the accuracy of sample results. The following methods have been used to collect data on television viewing in a sample household:

   a. The *diary method*. The household is asked to keep a diary of all programs watched and who watched them for a week and then mail in the diary at the end of the week.

   b. The *roster-recall method*. An interviewer shows the subject a list of programs for the preceding week and asks which programs were watched.

   c. The *telephone-coincidental method*. The household is telephoned at a specific time and asked if the television is on, which program is being watched, and who is watching it.

   d. The *automatic recorder method*. A device is attached to the t.v. and records what hours the set is on and to which channel it is turned. At the end of the week, this record is removed from the recorder.

For each method, discuss and record its advantages and disadvantages, especially any possible sources of error associated with each method. Method a. is most commonly used. Do you agree with that choice? Explain.

Searching for Misleading Data

Use a magazine(s) or newspaper section to find at least two examples of graphs or other data representations that you could argue are biased, misleading, or otherwise manipulative. Then find at least one that you would argue is not biased and has accurate results that are representative of a population. Record your reasoning for each of your samples, and be prepared to discuss your ideas.

Things to Know from Chapter 5

Words to Know

- box-and-whisker plot
- combination
- conditional probability
- correlation
- counting principles
- data (unpaired, paired)
- event (dependent, independent, mutually exclusive, random)
- expected value
- experiment (trial)
- fair
- frequency
- histogram
- interval
- mean (arithmetic average)
- median
- mode
- outcome (favorable, unfavorable)
- percentile
- permutation
- probability
- quartile (first, second, third, interquartile range)
- range
- relative frequency
- sample space
- sampling (representativeness)
- scatter plot
- standard deviation
- statistics
- weighted average

Concepts to Know

- what probability means, including conditional probability
- why probability is always a number, p, such that $0 \le p \le 1$
- what an event is (independent, dependent, mutually exclusive, random)
- what it means for an event to be fair
- why the counting principles are valid
- what are measures of central tendency and what can we learn from them
- what type of data are unpaired and what type of data are paired
- why one might use: box-and-whisker plot; histogram, scatter plot
- how sampling procedures can affect the data collected
- what representative means

Procedures to Know

- determining the probability of an event occurring
- determining the sample space for an event
- knowing when and how to use the counting principles
- finding mean, median, mode for a data set
- making box-and-whisker plot, histogram, scatter plot, stem-and-leaf plot
- interpreting information from a plot, table
- determining intervals, percentiles, quartiles for various plots

Exercises & More Problems

Exercises

1. How many elements are there in the sample space for each of the following experiments? Justify your answers.

   a. Tossing three dice
   b. Writing all three-letter "words" with a vowel as the middle letter
   c. Drawing two cards from a standard deck
   d. Answering five True-False questions

2. Suppose that a college dormitory keeps records by gender (F, M) and year in school (Fr., So., Jr., Sr., Grad). How many classifications are needed for the records? Why?

3. Suppose there are five sandwiches and four drinks that you like at a local fast-food establishment. From how many sandwich-drink combinations can you choose? What if there were also a choice of three desserts; then how many sandwich-drink-dessert combinations are there? Explain your reasoning.

4. Draw a tree diagram to represent the following sequence of experiments. Experiment I is performed twice. The three outcomes of experiment I are equally likely.

5. Draw a tree diagram to represent the following sequence of experiments. Experiment I is performed. Outcome *a* occurs with probability 0.3, and outcome *b* occurs with probability 0.7. Then experiment II is performed. Its outcome *c* occurs with probability 0.6, and its outcome *d* occurs with probability 0.4.

   a. What is the probability of having outcomes *a* and *d* occur?
   b. What is the probability of having outcomes *a* or *d* occur?

6. An experiment consists of selecting the last digit of a student identification number. Assume that each of the 10 digits is equally likely to appear as a last digit.

   a. List the sample space (i.e., the set of all possible outcomes).
   b. What is the probability that the digit is less than 5? Explain.
   c. What is the probability that the digit is odd? Explain.
   d. What is the probability that the digit is not 2? Explain.

7. A penny, a nickel, a dime, and a quarter are tossed. What is the probability of at least three heads?

8. An electric clock is stopped by a power failure. What is the probability that the second hand is stopped between the 5 and 6? Explain.

9. In how many different ways can the numbers on a single die be marked, with the only condition that the 1 and 6, the 2 and 5, and the 3 and 4 must be on opposite faces of the die?

10. Even though the March Hare, the Dormouse, and the Mad Hatter cried, "No room! No room!" when they saw Alice coming, it appears that there were actually nine places still available at the tea table. If we assume there were 12 places in all at the table, in how many ways can the four characters be seated around the table?

11. The Hindu god Siva has 10 arms and hands. In artists' renderings of Siva, he holds 10 items (e.g., a dagger, an arrow, a trident, a rope) in his hands. If these items can be exchanged from hand to hand, how many different variations in the appearance of Siva are possible?

12. A magician has a repertoire of eight magic tricks, of which he performs two on each show. In how many ways can he plan his program for three successive shows if:

    a. he does not want to perform the same trick on any two successive shows?
    b. he wants to use the same trick for the opening trick of each show?

13. The graph below shows how the value of a car depreciates each year. This graph will allow us to find the trade-in value of a car for each of five years. The percentages given in the graph are based on the selling price of the new car.

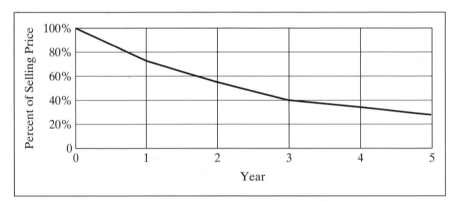

    a. What is the approximate trade-in value of a $12,000 car after one year? Explain.
    b. How much has a $20,000 car depreciated after five years? Explain.
    c. What is the approximate trade-in value of a $20,000 car after four years?
    d. Dani wants to trade in her car before it loses half its value. When should she do this? Explain.

The following table gives the percent of people 25 years or older in the United States and the number of years of school they have completed. Use this table to answer exercises 14–19.

Percent of Population Completing School

	0–4 years	5–7 years	8 years	9–11 years	12 years	13–15 years	16 or more years
1970	5.5	10.0	12.8	19.4	31.1	10.6	10.7
1980	3.6	6.7	8.0	15.3	34.6	15.7	16.2
1987	2.4	4.5	5.8	11.7	38.7	17.1	19.9

Source: *Statistical Abstracts of the United States*, 1989

14. In 1987, what percent of the people in the United States had at least a high school education? Explain.

15. In 1970, what percent of the people 25 and older had not completed a high school education? Explain.

16. If a random sample of 3,000 people was selected in 1987, approximately how many people in the sample would you expect to find had not completed a high school education?

17. In 1970, what percent of the people who started college had finished? Explain.

18. In 1987, what percent of the people who had started college had finished? Explain.

19. What trends can you observe from the data? What kind of sample might have produced this data?

20. An ice cream store offers 31 flavors. How many double-scoop cones are possible if the person eating the cone cares which scoop is on top?

21. An ice cream store offers 31 flavors. How many double-scoop cones are possible if the person eating the cone doesn't care which scoop is on top?

22. In how many different orders can a row of 10 people arrange themselves?

Critical Thinking

23. In bowling, how many different outcomes are possible on the first roll?

24. Make up a problem that involves the probability of independent events. Write the problem, and then solve it.

25. Create a problem involving conditional probability. Solve the problem you have created.

26. Find or create a realistic set of unpaired data with at least 10 data points.
    a. List them in a table, and then represent them with a box-and-whisker plot. Summarize the information provided in the plot, and give your interpretation of the data.
    b. Compare and contrast box-and-whisker plots, histograms, and scatter plots as ways to represent sets of data.

27. Suppose you play a game involving the roll of a fair die. The game has the following rules: (1) If you roll an even number, you receive twice the number of dollars as the number of dots on the die; (2) if you roll an odd number, you must pay twice the number of dollars as the number of dots on the die. Is this a fair game? Explain your answer. (You may want to compute the expected winnings from playing the game.)

28. You are cooking some spaghetti when suddenly one piece falls on the floor and breaks into three pieces. You friend, Henrietta, likes to bet. (In fact, she will bet on almost anything!) Henrietta takes a quick look at the three broken pieces of spaghetti on the floor and says, "I'll bet you $5 that I can form a triangle with those broken pieces of spaghetti." What is your chance of winning the bet?

29. During the past 10 years, the number of accidents on a busy highway on a Monday morning varies from 0 to 5. If the probabilities of the various numbers of accidents are given in the table, what is the expected number of accidents on a randomly selected Monday morning?

Number of accidents	0	1	2	3	4	5
Probability	0.62	0.15	0.10	0.08	0.02	0.03

30. The following are the amounts (rounded to the nearest dollar) paid by 25 students for textbooks during the fall term: 35, 42, 37, 60, 50, 42, 50, 16, 58, 39, 33, 39, 23, 53, 51, 48, 41, 49, 62, 40, 45, 37, 62, 30, 23.

   a. Make a histogram to represent the data.
   b. Make two interpretations about the data from the stem-and-leaf plot.
   c. Change the representation to a box-and-whisker plot. Find the median, first quartile, and third quartile values.
   d. Explain why the value of the third quartile is not a whole number.

31. The distortion factor (D.F.) is a measure of the extent to which a graph misrepresents a set of data. The distortion is measured by the following ratio: $D.F. = \dfrac{G}{D}$, where G is the maximum change in the graph (as a ratio) and D is the maximum change in the data (as a ratio). For example, the graph below distorts the actual data for average length of the school year in various countries. For this graph, $G = 1.875$ in./0.5625 in. $= 3.3333$ and $D = 245$ days/178 days $= 1.3764$. Thus, $D.F. = 3.3333/1.3764 = 2.4217$.

   a. Would the distortion factor be larger or smaller if the y-axis started at 0?
   b. What does a distortion factor of 1 represent?
   c. How should the graph be changed to eliminate any distortion?

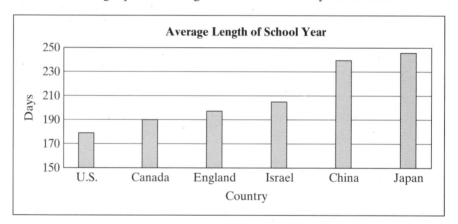

Extending the Activity

32. If the spinner shown is spun, find the probabilities of obtaining each of the following:

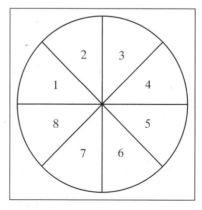

   a. P (factor of 35)—remember that this means the probability of getting a factor of 35
   b. P (multiple of 3)
   c. P (even number)

d. P (6 or 2)
  e. P (11)
  f. P (composite number)
  g. P (neither a prime nor a composite)

33. Paul, Norm, Lois, Kathy, and Diana are in a round-robin racquetball tournament. Individuals of the same sex have equal probabilities of winning, but each man is twice as likely to win as any woman.

    a. What is the probability that a woman wins the tournament?
    b. If Paul and Diana are married, what is the probability that one of them wins the tournament?

34. Four AAA batteries are chosen at random from a set of 16 batteries of which six are defective. Find the probability that:

    a. Exactly one of the four batteries selected is defective.
    b. None of the four is defective.
    c. At least one is defective.

35. From a pool of 20 smokers and 10 nonsmokers, a group of 10 is to be chosen for a medical study.

    a. In how many ways can this be done if it does not matter how many in the group are smokers and how many are nonsmokers?
    b. In how many ways can this be done if exactly five members of the group must be smokers and five must be nonsmokers?

36. An experiment consists of rolling a single die and observing the number showing on the top face. Give the sample space and the probability distribution for the experiment.

37. An experiment consists of rolling a pair of dice and observing the product of the numbers showing on the top faces. Give the sample space and the probability distribution for the experiment.

38. A box of 200 transistors contains 10 defectives. An inspector draws five transistors at random and determines the number of defectives. The box is rejected if one or more of those selected are defective. Find the probability that the box is rejected.

39. Jamal Q. Student forgets to set his alarm with a probability of 0.2. If he sets the alarm, it rings with a probability of 0.9. If the alarm rings, it will wake him on time for his 8:30 A.M. class with a probability of 0.8. If the alarm does not ring, he wakes in time for his 8:30 A.M. class with a probability of 0.3.

    a. Draw the tree diagram for this situation, and give the appropriate probabilities on the branches.
    b. Find the probability that Jamal wakes in time for his 8:30 A.M. class tomorrow.

40. A pizza shop conducted a survey to find out which ingredients (extra cheese, green peppers, or mushrooms) its customers wanted on their pizzas. Of the 100 respondents, 60 wanted extra cheese, 25 wanted green peppers, and 50 wanted mushrooms. In addition, 10 people wanted extra cheese and green peppers, 15 wanted green peppers and mushrooms, 25 wanted extra cheese and mushrooms. Finally, five people wanted extra cheese, green peppers and mushrooms.

    a. Draw a Venn diagram depicting the results of this survey. Include in it the number of people that fall into each of its regions.

b. How many people wanted only mushrooms?
c. How many people wanted green peppers and mushrooms but not extra cheese?

41. An urn contains 55 red and 45 green balls. A player draws a ball at random. If the green ball is drawn, the player wins $5 and the game ends. If a red ball is drawn, the player draws again from the remaining balls in the urn and wins $1 if a green ball is drawn. The game ends after two draws. [Assume that there are no more than two draws.]

   a. Draw the tree diagram for this game, and give the appropriate probabilities for each branch.
   b. Make a probability distribution for the earnings and the probabilities.
   c. Find the expected winnings for the game.

42. A survey of 100 college faculty who exercise regularly found that 45 jog, 30 swim, 20 cycle, six jog and swim, one jogs and cycles, five swim and cycle, and one does all three. How many of the faculty members do not do any of these three activities? How many just jog?

43. An experiment consists of tossing a coin eight times and observing the sequence of heads and tails.

   a. How many different outcomes are possible?
   b. How many different outcomes have exactly three heads?
   c. How many different outcomes have at least two heads?
   d. How many different outcomes have four heads or five heads?
   e. Explain a different way you could figure out an answer for c.

44. A letter is selected at random from the word *MISSISSIPPI*.

   a. What is the sample space for this experiment?
   b. Describe the event "the letter chosen is a vowel" as a subset of the sample space.

45. A town is selected at random from the 60 towns in a particular geographical region.

   a. What is the probability that it is one of the nine largest cities in this region?
   b. What is the probability that it is one of the 12 towns in the southern part of this region?

46. An exam contains five "true-or-false" questions. What is the probability that a student guessing at the answers will answer three or more correctly?

47. Two friends are playing a new game involving flipping two standard coins. Eric wins a point if the coins match (i.e., if the coins come up with two heads or two tails). Shelley wins a point if the coins do not match. Answer each of the following questions:

   a. Who is more likely to win, Eric or Shelley? Explain your answer.
   b. Is this a fair game? Explain your answer.

48. Suppose you have a regular deck of 52 playing cards.

   a. What is the probability of drawing a heart from a deck of 52 cards?
   b. What is the probability of drawing two hearts in a row if the first card is replaced before the second card is drawn?
   c. What is the probability of drawing two hearts in a row if the first card is not replaced before the second card is drawn?

49. In a family with three children, what is the probability that at least one child is a girl? What assumptions did you have to make to answer this question?

50. What is the largest number of pieces into which you can cut a pizza with ten slices? (The slices may be of different sizes and shapes, and you may assume that slices are not stacked on top of each other.)

51. How many chess matches are needed in a round-robin tournament with 10 players (where each player in the tournament plays against each of the other players)?

52. How many chess matches are needed in an elimination tournament with 10 players (where a player is eliminated if he loses)?

53. How many handshakes does it take for a roomful of 10 strangers to introduce themselves to each other?

54. Find or create a realistic set of paired data with at least 10 data points. List them in a table.
    a. Summarize the information in your table in a bar graph.
    b. Summarize the information in your table in a scatter plot.
    c. Is there any information provided by the bar graph that you cannot get from the scatter plot?
    d. Is there any information provided by the scatter plot that you cannot get from the bar graph?

55. Read the discussion of distortion factor in Exercise #31.
    a. Find a graph in a newspaper or magazine that displays numerical data and figure its distortion factor. [*USA Today* is a good source of graphs.]
    b. Redraw the graph so that the distortion factor is 1.

Analyze the data in the table below. Use this table to complete exercises 56–61.

Final Medal Standings for the Top 20 Countries—1988 Olympics

Country	Number of Medals
USSR	132
East Germany	102
United States	94
West Germany	40
Bulgaria	35
South Korea	33
China	28
Romania	24
Great Britain	24
Hungary	23
France	16
Poland	16
Italy	14
Japan	14
Australia	14
New Zealand	13
Yugoslavia	12
Sweden	11
Canada	10
Kenya	9

56. Represent the data in a histogram. What scales will you use for the intervals and frequency?

57. List two interpretations that you can make by looking at the histogram.

58. Represent the data in a box-and-whisker plot. Determine the following values: median, first quartile, third quartile.

59. An *outlier* is a value that is widely separated from the rest of a group of data. Use the following rule to determine which, if any, of the data points in your list are outliers: An outlier is any value that is more than 1.5 interquartile ranges above the third-quartile value or more than 1.5 interquartile ranges below the first-quartile value. The interquartile range is the difference between the third-quartile and first-quartile values.

60. Eliminate any outliers from the data set above, and determine the following values: median, first quartile, third quartile.

61. What number represents the best answer to the question, "What is the average number of medals won by the top 20 countries?" Explain.

Writing/Discussing

62. Think about events that may occur in your life.

    a. List three events that are certain to occur, and explain why they are certain.

    b. List three events that are impossible to occur (i.e., cannot occur), and explain why they are impossible.

    c. List three events that are highly likely, and explain why they are highly likely.

    d. List three events that are unlikely, and explain why they are unlikely.

63. Think about the Spinner and What's in the Bag? experiments in Activity 5.1. You made predictions and/or obtained experimental results. How close do you think your predictions were? What is a reasonable definition of probability?

64. Consider the Dice experiment of Activity 5.2. Describe a general method for finding out the probability of one event or another event occurring at the same time.

65. Explain the relationship between an outcome, an event, and the sample space for a particular experiment.

66. James tossed a coin 15 times and claimed that it turned up heads every time.

    a. Is this possible? Explain your answer.

    b. Is it likely? Explain your answer.

    c. What is likely to happen on the 16th toss of the coin? Explain your answer.

67. Give a reasonable argument why probabilities are never less than 0 and never greater than 1.

68. Write a general statement explaining the process of determining combinations.

69. Formulate and solve a problem analogous to #27.

70. The article shown on page 221 appeared in *USA Today*. Read the article, and then create a graph or a chart that you feel would display the information conveyed as well or better than just using words.

A's in tough courses beat SATs, ACTs
How to get into college of your choice, 4D;
Wednesday: Financial aid

By Pat Ordovensky USA TODAY

High school grades and the courses in which they're earned, are the most important factors in deciding who gets into college, say USA admissions directors. Scores on college admissions exams—the SAT and ACT—rank third in importance among items on a student's application. Choosier colleges say a student's essay has greater weight than test scores.

So finds a USA TODAY survey of 472 admissions directors, selected randomly from four-year colleges.

The findings echo recent public statements by admissions officers that the SAT and ACT are losing value in predicting college success—that A's in tough high school courses are more relevant.

"Almost everyone keys on academic progress over 3 1/2 years" in deciding which applicants to admit, says Richard Steele, Duke University admissions director. "The combination of high grades and a quality high school program is the best predictor of success."

For the survey, USA TODAY talked to large private universities with well-known names and small state colleges that specialize in training teachers. Among them are 27 highly selective schools—from Harvard and Princeton to Stanford and Cal Tech—that accept an average of only 31 percent of their applicants. We found:

- Almost 9 of 10 (88 percent) say grades are "very important" in admissions; 67 percent say rigor of high school courses; 52 say SAT/ACT scores; 45 percent say class rank.
- 58 percent of all schools and 52 percent of the choosiest say grades are the "most important" of all factors. Second is tough high school courses, mentioned by 16 percent of all schools, 28 percent of the choosiest.
- SAT and ACT scores are "most important" at 7 percent of the campuses but none of the selective schools.
- 17 percent of all schools but 60 percent of the choosiest say an applicant's essay is a "very important" factor.
- All schools responding have an average $1.5 million of their own money to help students who can't afford the tuition. At the selective schools, an average of $9 million is available.
- 89 percent of the selective schools used a waiting list this year for qualified applicants who didn't make the final cut. Average number of names on the list: 396.

Brown University admissions director James Rogers, an author of the new book 50 College Admissions Directors Speak to Parents, writes that experienced admissions officers "can review a transcript and predict the student's test scores."

And although women consistently score lower than men on SAT/ACT, 52 percent of this year's freshmen are women.

Other factors—recommendations, interviews—enter in, says Steele, because with talented applicants, "you don't eliminate large numbers if you stop at academic performance."

On less selective campuses, grades carry even more weight.

At 15 percent of the colleges responding, because of state law or school policy, SAT/ACT scores are deciding factors.

Three of every 10 schools surveyed (29 percent) say they're required to accept students meeting certain academic criteria. More than half (54 percent) say criteria include test scores.

The University of Arizona, for example, is required to accept any Arizona high school graduate who meets four requirements: a 2.5 (C-plus) grade average, 11 college prep courses, standing in top half of class, a 930 SAT score or 21 on the ACT.

Forty percent of Arizona's students are from out-of-state, says admissions director Jerry Lucido, but they need at least a 3.0 (B) high school average, a 1010 on the SAT or 23 ACT score.

At Michigan, North Carolina, Vermont and other public universities where more than twice as many non-residents apply as can be accepted, officials also say standards are much tougher for students who don't live in their states.

Among other factors considered important in the admission processes, the USA TODAY survey finds:

- **Minority status:** Almost two-thirds (62 percent) of all schools and 96 percent of selective colleges say they're actively recruiting minority students who meet academic standards. That means minority status makes a difference between otherwise equal applications.

Of schools in the survey, this year's freshman class has an average 9 percent minorities—3 percent black.

- **Recommendations:** 70 percent of all schools, 96 percent of the most selective, say recommendations from teachers and guidance counselors are important. More than half (52 percent) of the choosy schools say they're very important.

Other recommendations, from the community leaders, clergymen and such, are important to 52 percent of all schools. 75 percent of the choosiest.

But "choose your references carefully," says Steele. "More is not better.... Last year, an applicant gave us 23 unsolicited letters of recommendation. We were not impressed."

- **Interviews:** Many highly selective schools use alumni across the USA to interview applicants in their area. At Duke, alumni reports weigh the same as teacher's recommendations.

Almost half (45 percent) of the surveyed schools and 64 percent of the choosiest say an interview is important.

- **Essay:** It's important to 92 percent of the selective schools but only half (50 percent) of all schools.

It offers a clue to a student's "quality of thinking," says Steele. "Few kids can write their way in, but it can help."

Errors—factual and grammatical—hurt. At Boston University a few years ago, an applicant was rejected because his essay had too much Whiteout.

- **Geography:** Most selective schools pride themselves on having students from all parts of the USA. A student from Montana applying to Harvard, for example, gets an edge over a New Englander.

At Duke, "we don't get many from the Dakotas," says Steele. "We snap to when one walks into the office."

More typical of all schools is Arizona where, Lucido says. "We're not shooting for geographic balance. We give no geographic preference."

- **Alumni ties:** Admission directors at selective schools say candidly that children of Alumni get special preference and children of generous alumni are even more special.

An alumni relationship is important to 44 percent of all schools, 80 percent of the most selective.

At Duke, 20 percent of all applicants are accepted. But for "alumni-connected" candidates the rate is over 40 percent.

Copyright 1988, *USA TODAY*.

CHAPTER SIX

Fraction Models & Operations

CHAPTER OVERVIEW

Many real-world situations require the use of fractions. The primary purpose of this chapter is to extend your sense of fractions and your understanding of operations on fractions. Because a deep understanding of fractions is essential for anyone who will teach fraction concepts and procedures to children, the first several activities involve considerations of various ways to interpret and model (that is, represent) fractions. Then, in addition to activities aimed at helping you develop better fraction sense, the chapter moves on to investigations involving fraction computations and everyday applications of fractions.

BIG MATHEMATICAL IDEAS

Problem-solving strategies, conjecturing, verifying, decomposing, generalizing, using language and symbolism, mathematical structure

NCTM PRINCIPLES & STANDARDS LINKS

Number and Operation; Problem Solving; Reasoning; Communication; Connections; Representation

Activity **6.1** Introducing the Region Model
6.2 Introducing the Linear Model
6.3 Introducing the Set Model
6.4 Exploring Fraction Ideas Through the Region Model
6.5 Fractions on the Square: A Game Using the Region Model
6.6 Fraction Puzzles Using the Region Model
6.7 Looking for Patterns with the Linear Model
6.8 Exploring Fraction Ideas Through the Linear Model
6.9 Exploring the Density of the Set of Real Numbers
6.10 Exploring Fraction Ideas Through the Set Model
6.11 Solving Problems Using the Set Model
6.12 Classifying Problems by Operation: Revisiting Activity 3.3
6.13 Illustrating Operations with Region, Linear, and Set Models
6.14 Developing Fraction Sense with Linear Models
6.15 Using the Region Model to Illustrate Multiplication
6.16 Using the Region Model to Illustrate Division

ACTIVITY 6.1 Introducing the Region Model

FYI Topics in the Student Resource Handbook

6.1 **Fractions**

For this activity, your instructor will give you five envelopes with three pieces of a square in each envelope. Your task is to reassemble the five squares following these rules:

1. No talking is allowed.
2. You may *give* a piece to another student, but the only way you can obtain a piece is by someone giving you one of his or hers.
3. After all the squares are completed, assign a fraction value to each piece. An assembled square is one unit. Talking is allowed during this time. You should not write on any of the pieces.
4. Record the value of each piece below, and explain how you obtained this value.

#1 =   #2 =

#3 =   #4 =

#5 =   #6 =

#7 =   #8 =

#9 =   #10 =

#11 =   #12 =

ACTIVITY 6.2 Introducing the Linear Model

FYI Topics in the Student Resource Handbook

6.1 **Fractions**

Prove, using Cuisenaire Rods, that the yellow rod is half as long as the orange. This relationship can be written as 1/2 o = y. Find the other pairs of halves, and record them. Then do the same for thirds, fourths, fifths, and so on up to tenths. Record your findings. Then write the expressions the other way; for instance, 2y = 1o. Find at least 4 pairs for each fractional relationship.

ACTIVITY 6.3 Introducing the Set Model

FYI Topics in the Student Resource Handbook
6.1 Fractions

In an adult condominium complex, 2/3 of the men are married to 3/5 of the women. What part of the residents are married?

ACTIVITY 6.4 Exploring Fraction Ideas Through the Region Model

FYI Topics in the Student Resource Handbook
6.2 Equivalent Fractions

1. Cut out the strips and other shapes in Appendix A on the page your instructor indicates. Do *not* cut out the shapes on page 228.

2. Take the two strips of paper, and tear one into two pieces and one into three pieces.

3. Compare your strips with your group members' strips, and note the similarities and differences.

4. What directions need to be given so that each person ends up with strips equal in size after tearing?

5. Use the squares, rectangles, hexagons, and circles that you cut out to divide the shapes into halves and thirds, following the steps below. In each case, use a cutout shape for folding, cutting, or drawing, and then record your work on an identical shape on page 228.
    a. First, divide the shape by folding it along lines of symmetry into congruent pieces.
    b. Second, divide the shape by cutting (not on lines of symmetry) into congruent pieces. Verify that you have divided the shape into halves (or thirds) by placing the pieces on top of each other.
    c. Third, divide the shape by drawing line segments that separate it into equal-area, noncongruent parts.

6. How many ways were you able to fold a square in half? A rectangle? A hexagon? A circle? A square in thirds? A rectangle in thirds? A hexagon in thirds? A circle in thirds?

7. A square can be folded in half many different ways. Find ways that are different from the ways you folded it. Can you make a generalization about this folding process? Will it work for rectangles, hexagons, and circles also?

ACTIVITY 6.5 Fractions on the Square: A Game Using the Region Model

Topics in the Student Resource Handbook
6.2 Equivalent Fractions

Play this game with another group member. Each player, in turn, throws two dice with fractional values on the faces and decides which is the larger of the two fractions appearing on the dice (or whether the fractions are equivalent). If the player is correct, he or she can shade in, on one of the squares, a region equivalent to the fraction. The winner is the first player to shade in all squares on her or his sheet completely.

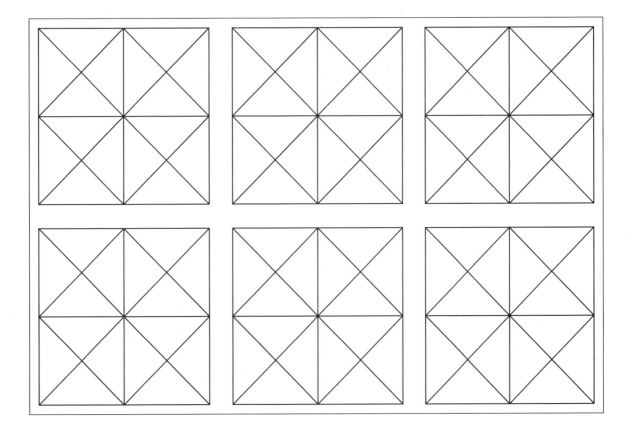

ACTIVITY 6.6 Fraction Puzzles Using the Region Model

FYI Topics in the Student Resource Handbook
6.2 Equivalent Fractions

Build the following figures using pattern blocks, and then sketch your figure.

1. Build a triangle that is 1/3 green and 2/3 red.

2. Build a triangle that is 2/3 red, 1/9 green, and 2/9 blue.

3. Build a parallelogram that is 3/4 blue and 1/4 green.

4. Build a parallelogram that is 2/3 blue and 1/3 green.

5. Build a trapezoid that is 1/2 red and 1/2 blue.

ACTIVITY 6.7 Looking for Patterns with the Linear Model

FYI Topics in the Student Resource Handbook
6.2 Equivalent Fractions

With Cuisenaire rods, there are four different ways to make arrangements, called trains, equivalent in length to the light-green rod: light green, red-white, white-red, and white-white-white.

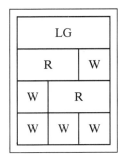

Do the same for the other rods. Find a pattern that allows you to predict how many different ways there are to make trains for any length rod.

ACTIVITY 6.8 Exploring Fraction Ideas Through the Linear Model

FYI Topics in the Student Resource Handbook
6.2 Equivalent Fractions

Cut out and use strips #1, 2, 3, and 4 in Appendix A on the page that your instructor indicates.

1. Suppose that strip #1 is one unit long. Use it to model 2/3. Illustrate the process you used, and shade the answer.

2. Do the same for 3/4 and 3/2, using #1 strips from more than one group member.

3. Make a number line using strip #1 as 1 unit, and place the numbers 2/3, 3/4, and 3/2 in the appropriate locations.

4. If strip #1 represents 1 unit, what does strip #2 represent? strip #3? strip #4?

5. Model 2/3 with strip #2. Note that this is 2/3 unit, not 2/3 of strip #2. Illustrate with a diagram the process you used, and shade the answer. Represent the process with mathematical symbols.

6. Model 3/4 with a strip other than #1. Illustrate the process you used, and shade the answer. Represent the process with mathematical symbols.

7. Model 4/5 with two different strips. Illustrate the process you used, and shade the answer. Represent the process with mathematical symbols.

8. Make a generalization concerning the following: Given a fraction, how can you represent it two different ways with strips of paper?

9. If you have a 1-pound bar of chocolate, how much is 3/4 of it?

10. If you have a 3-pound bar of chocolate, how much is 1/4 of it?

11. Explain why your answers are the same.

12. For each of the following, model the operation with a strip of paper, illustrate with a diagram the process you used, and shade your answer.

    a. 3 × 1/4
    b. 1/4 × 3
    c. 1/3 × 1/4
    d. 1/4 × 1/3

ACTIVITY 6.9 Exploring the Density of the Set of Real Numbers

Topics in the Student Resource Handbook

7.11 Rational Numbers
7.12 Looking at Rational Numbers as the Set of Rational Numbers

1. Use two pieces of paper taped together end-to-end to make a number line. The place where the two pages meet can serve as the origin. Using strip #1 as length 1, mark on the number line all the integers that you can. Use one of the strips of paper to mark all the halves. Justify the process that you used to find these. Do the same for thirds and fourths.

2. Is it possible to use this same procedure to find other fractions on the number line?

3. If you were asked to mark all the sixths, what steps would you take to do this using strips of paper?

4. Describe where you would place 1/6 on the number line and why you would place it there.

5. Is it possible to use a certain number of sixths to represent the same number as 1/2? If so, how many?

6. Is it possible to use a certain number of thirds to represent the same number as 1/2? If so, how many? Is it possible to use a certain number of fifths to represent the same number as 1/2? If so, how many?

7. If you have a fraction of the form 1/*n*, when will it be possible to use a certain number of 1/*n*'s to represent the same number as 1/2?

8. How many sixths do you need to represent the number 2? How many thirds? How many fourths?

9. Do you think that every number has more than one representation? Justify your reasoning.

10. Now look at your number line. Consider the part from 1 to 3/2. Is there a number between 1 and 3/2? What is it?

11. Now, look at the part of the number line from 3/2 to 5/3. Is there a number between these two numbers? What is it? Can you find another number between 3/2 and 5/3? How many such numbers can you find?

12. Make a conjecture concerning the existence of a number between any two numbers.

ACTIVITY 6.10 Exploring Fraction Ideas Through the Set Model

FYI Topics in the Student Resource Handbook
6.2 Equivalent Fractions

Using 12 color tiles, model the following and illustrate your answer.

1. $\dfrac{2}{3}$

2. $\dfrac{3}{4}$

3. $\dfrac{1}{2}$

4. $\dfrac{5}{6}$

5. $\dfrac{1}{2} \times \dfrac{2}{3}$

6. $\dfrac{3}{4} \times \dfrac{1}{3}$

Can you illustrate division of fractions with these tiles? Why or why not?

ACTIVITY 6.11 Solving Problems Using the Set Model

FYI Topics in the Student Resource Handbook
6.2 Equivalent Fractions

1. I'm thinking of a number. One-half is a third of my number. What is my number?

2. A balance scale was in perfect balance when Horace placed a box of candy on one pan of the balance and 3/4 of the same-sized candy box together with a 3/4-pound weight on the other pan. How much did the full box of candy weigh?

3. Every time Agnes goes to a flea market, she buys another antique quilt. She has quite a collection of quilts, but she is reluctant to admit exactly how many she has.
    Alexander asked her directly, "How many quilts do you have now?"
    "Oh, I don't really have that many," Agnes replied evasively. "Actually, I have three-quarters of their number plus 3/4 of a quilt."
Alexander thought she must be kidding, but she was actually challenging him with a problem. How many quilts does Agnes have?

4. Samuel was riding in the back seat of the station wagon on the way home after a long and tiring day at the beach. He fell asleep halfway home. He didn't wake up until he still had half as far to go as he had already gone while asleep. How much of the entire trip home was Samuel asleep?

240 Chapter 6 Fraction Models & Operations

ACTIVITY 6.12 Classifying Problems by Operation: Revisiting Activity 3.3

FYI Topics in the Student Resource Handbook

7.13 **Adding Rational Numbers**
7.14 **Subtracting Rational Numbers**
7.15 **Multiplying Rational Numbers**
7.16 **Dividing Rational Numbers**

Recall Activity 3.3 where you classified word problems into different categories for each of the four operations. In each of the blanks below, write the various classifications for each operation. For example, for subtraction there are the classifications of comparison, take away, and missing addend. Make up one word problem using fractions for each classification.

	Type I	Type II	Type III
Addition	_____	_____	_____
Subtraction	_____	_____	_____
Multiplication	_____	_____	_____
Division	_____	_____	_____

ACTIVITY 6.13 Illustrating Operations with Region, Linear, and Set Models

FYI Topics in the Student Resource Handbook

7.13 Adding Rational Numbers
7.14 Subtracting Rational Numbers
7.15 Multiplying Rational Numbers
7.16 Dividing Rational Numbers

Translate these into mathematical symbols, illustrate the problem, and solve it.

1. Two-ninths of it is one-ninth. What is it?

2. One-fifth of it is one-third. What is it?

3. One-fourth of it is one-sixth. What is it?

Fold a strip of paper to illustrate the relationship in the diagram, and record with mathematical symbols the action that is taking place.

4.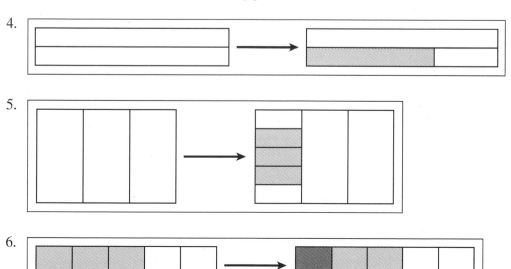

5.

6.

Chapter 6 Fraction Models & Operations

Model these with paper folding, and draw an illustration.

7. 1/2 of 2/7

8. 2/7 of 1/2

Draw an illustration to represent the problem and solve.

9. There are four bottles. Each bottle contains 2/5 of a liter. How many liters are there altogether in the bottles?

10. There are three bags of marbles, each with the same number of marbles. If 2/3 of the marbles are taken from each bag, what part of each bag remains? How many full bags of marbles could be made from the remaining marbles?

11. If a school cafeteria makes 44 pints of juice for lunch each day and each lunch period group uses 11 pints, how many lunch period groups are there?

12. Now suppose that all the juice needs to be put into containers holding only 1/2 of a pint. How many containers will be needed?

Illustrate these problems and solve.

13. How many halves are contained in 3 units?

14. How many thirds are in 23? [Hint: Illustrate a simpler problem, and use your conclusion to solve this problem.]

15. How many times is 2/7 contained in 4/7?

16. How many times is 3/11 contained in 9/11?

17. How many times is 4/11 contained in 9/11?

18. What conclusions can you make from the three previous problems?

19. How many times is 2/5 contained in 3/7? [Try to use what you learned in the above problems and the concept of fractions to explain the solution to this problem.]

ACTIVITY 6.14 Developing Fraction Sense with Linear Models

FYI Topics in the Student Resource Handbook

7.13 Adding Rational Numbers
7.14 Subtracting Rational Numbers
7.15 Multiplying Rational Numbers
7.16 Dividing Rational Numbers

1. *a, b, c* represent whole numbers different from 0 and $a > b > c$. What can you say about:

   a. *a/b*   b. *b/a*   c. *b/c*

   d. *c/b*   e. *a/c*   f. *c/a*

2. Which is larger:

   a. *a/c* or *b/c*?   b. *a/b* or *b/b*?

   c. *a/b* or *a/c*?

3.

   a. If the fractions represented by the points D and E are multiplied, what point on the number line best represents the product?

   b. If the fractions represented by the points C and D are operated on as C/D, what point on the number line best represents the quotient?

   c. If the fractions represented by the points B and F are multiplied, what point on the number line best represents the product?

   d. Suppose 20 is multiplied by the number represented by E on the number line. Estimate the product.

   e. Suppose 20 is divided by the number represented by E on the number line. Estimate the quotient.

4.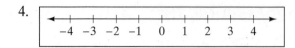

   a. Indicate, by using A, where 1/2 + 3/4 is on the number line.
   b. Indicate, by using B, where 1/2 − 3/4 is on the number line.
   c. Indicate, by using C, where 1/2 · 3/4 is on the number line.
   d. Indicate, by using D, where 1/2 ÷ 3/4 is on the number line.

5. Use the 10 digits to make the following numbers. Make sure that you use each digit exactly once in every number.

   a. The smallest positive fraction you can make using two distinct digits, three distinct digits, four distinct digits, ... , 10 distinct digits

   b. The largest fraction, not equal to a whole number, you can make using two distinct digits, three distinct digits, four distinct digits, ... , 10 distinct digits

   c. A fraction very close to, but not equal to, 2

   d. A number that is between 1/4 and 3/4

   e. A number that is between 7/8 and 1

   f. A number that is half as large as 1/6

   g. The number that is closest, but not equal to, 9/10 using four digits

   h. An estimate of 1/3 of $26.25

   i. An estimate of 1/4 of $27.50

ACTIVITY 6.15 Using the Region Model to Illustrate Multiplication

FYI Topics in the Student Resource Handbook
7.15 Multiplying Rational Numbers

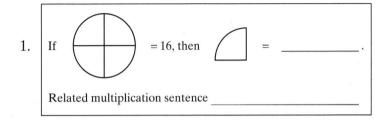

1. If ⊕ = 16, then ◔ = _____.

   Related multiplication sentence _____

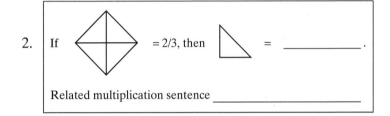

2. If ◇ = 2/3, then △ = _____.

   Related multiplication sentence _____

3. If [▭▭▭] = 3, then ▭ = _____.

   Related multiplication sentence _____

4. If ▱ = 12, then ╱ = _____.

   Related multiplication sentence _____

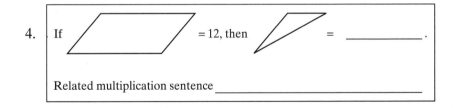

5. If ⬡ = 9, then ▱ = _____.

   Related multiplication sentence _____

ACTIVITY 6.16 Using the Region Model to Illustrate Division

FYI Topics in the Student Resource Handbook
7.16 **Dividing Rationale Numbers**

Recall from Activity 6.4 the generalizations you made regarding dividing figures in half in infinitely many ways. For this activity, your group will be investigating dividing figures into equal-area, noncongruent pieces.

1. Find a method of dividing this rectangle into two noncongruent pieces with equal area.

Make a generalization about your method of division.

2. Now generalize the division for 3, 4, ..., n noncongruent pieces with equal area.

3. A square-topped cake that is a rectangular solid is frosted on all faces except the bottom. Cut it into five pieces so that each person gets the same amount of cake and the same amount of frosting. All cuts must be perpendicular to the top of the cake. Each person must get their cake in one piece.

Things to Know from Chapter 6

Words to Know

- density
- equivalent fractions
- fraction
- models of fractions (region, linear, set)
- number line
- simplest terms

Concepts to Know

- what is a fraction
- what is meant by equivalent fractions
- what the region, linear, and set models of fractions represent
- what it means to put a fraction in simplest terms
- what it means to say that the real number line is dense

Procedures to Know

- recognizing equivalent fractions
- modeling fractions with region, linear, and set models
- modeling operations with fractions
- simplifying fractions
- classifying word problems by operation category
- finding a number between two other numbers

Exercises & More Problems

Exercises

1. Combine the first 2/3 of "ten" with the last 2/3 of "Sam."

2. Combine the first 1/2 of "blue" with the last 3/4 of "send."

3. Combine the first 1/3 of "materials" with the last 3/4 of "this" and the first 1/4 of "fundamentals."

4. Take the first 2/3 of the first 1/3 of "different." Then, take the second letter of the last 1/3 of "different." Take the second letter of the first 1/3 of "different." Now, take the third letter of the first 4/9 of "challenge." Take the first 1/2 of the last 1/2 of the first 4/9 of "challenge." Finally, take the first 1/3 of the last 1/3 of "challenge." Reorder all the letters you have obtained to form a familiar word.

5. Make up two word puzzles, similar to the one in #4, that use multiplication of fraction ideas.

6. Shade in the appropriate region(s), or place at the appropriate place on the number line so that the result shows the given fraction:

   a. $\dfrac{3}{4}$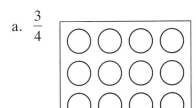

   b. $\dfrac{2}{5}$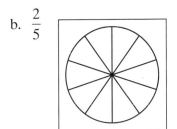

   c. $\dfrac{5}{7}$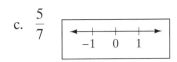

7. a. Assume that each square below represents 1. What fraction is shaded in each? Write a multiplication sentence that would make sense in each picture below.

   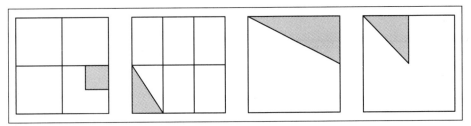

   b. Make up two other pictures like the ones above, and write multiplication sentences for them.

8. Suppose there is a contest to decide on a design for the flag of a new republic. Draw four different designs that have the rectangular flag 1/4 blue, 1/4 red, and 1/2 yellow.

9. Write two fractions for each of the following that equal the given fraction:

   a. $\dfrac{3}{9}$
   b. $\dfrac{-3}{5}$
   c. $\dfrac{0}{2}$

10. Draw an area model to show that $\dfrac{2}{3} = \dfrac{6}{9}$

11. Compute each of the following:

    a. 2/7 + 8/11
    b. −5/13 + 4/7
    c. 2/5 − 7/9

12. Model the following situations by drawing an illustration:

    a. 1/3 of 3/5
    b. 2/9 of 1/2

13. Determine which fraction does not equal the other two by reducing to simplest form:

    a. $\dfrac{7}{13}, \dfrac{28}{52}, \dfrac{35}{78}$

    b. $\dfrac{-32}{63}, \dfrac{4}{-9}, \dfrac{-64}{144}$

    c. $\dfrac{17}{21}, \dfrac{34}{63}, \dfrac{102}{126}$

14. Arrange each of the following in increasing order:

    a. $\dfrac{7}{11}, \dfrac{7}{14}, \dfrac{7}{9}$

    b. $\dfrac{-3}{5}, \dfrac{-7}{8}, \dfrac{-8}{19}$

15. Find two rational numbers that are between the following two fractions:

    a. $\dfrac{7}{9}$ and $\dfrac{8}{9}$

    b. $\dfrac{-9}{14}$ and $\dfrac{-8}{14}$

    c. $\dfrac{13}{19}$ and $\dfrac{8}{11}$

16. Compute each of the following (simplify when necessary):

    a. $\dfrac{25}{42} \cdot \dfrac{28}{50}$

    b. $7 \div \dfrac{1}{2}$

    c. $\dfrac{\frac{7}{9}}{\frac{4}{18}}$

17. On the same number line determine where the following expressions should be placed:

    a. 3/2 + 3/4   b. 3/2 − 3/4   c. 3/2 × 3/4   d. 3/2 ÷ 3/4

18. Write a comparison subtraction problem involving the fractions 1/3 and 1/4, and solve the problem.

19. Illustrate how you could model, with a strip of paper, the quotient 0.2 ÷ 0.6. Explain.

20. Illustrate 0.9 ÷ 0.6 by drawing a model of base-ten blocks. Explain your model and solution.

Critical Thinking

21. Under what conditions would one positive rational number with a greater numerator than another rational number be greater in value?

22. Suppose you are given a number. If one-fourth is two-thirds of that number, what is the number?

23. Illustrate the following situation, and use that to solve the problem: Two-thirds of a number is three-fifths. What is the number?

24. Illustrate how many times is 4/9 contained in 6/9.

25. What might be a reasonable estimate for 1/6 of 82? Explain how you determined that estimate.

26. Is taking one-third of a number the same as dividing the number by one-third? Explain.

27. Diana bought a piece of cloth 48 inches wide and 1 yard long. It cost $10. She cut off one-third and used it to make a tablecloth. From the remaining material, she used a piece that was 12 inches wide and 12 inches long to make a scarf and a piece 2 feet by 1.5 feet to make the cover for a throw pillow. Her sister saw Diana's sewing efforts and said she really liked the material. She wanted to buy what was left to do some sewing of

her own. Diana was willing to sell the leftover material for what she had paid for it. How much should she charge her sister? Find three different methods of doing this problem. Use paper folding for at least one of your methods.

28. After a cyclist has gone 2/3 of her route, she gets a puncture in a tire. Finishing on foot, she spends twice as long walking as she did riding. How many times as fast does she ride as walk?

29. Moses can eat seven slices of pizzas in three hours, and Ryan can eat nine slices of pizzas in seven hours. How many pizzas can Moses and Ryan eat together in five hours? Justify your answer.

30. Five maids can clean seven houses in three days. How long would it take eight maids to clean 20 houses if all the work was done at the same rate?

31. If you have a fraction of the form $1/n$, when will it be possible to use a certain whole number of $1/n$'s to represent the same number as 1/2? Explain.

32. Adam put 1/2 of a cake in the freezer. Of the remaining half of the cake, Adam ate 1/5, and his dog ate the rest. What part of the whole cake did his dog eat?

   a. 1/2 − 1/5
   b. 1/2 · 1/5
   c. 1 − (1/2 · 1/5)
   d. 1/2 − 4/5
   e. none of these

33. Rebekah is making bookmarks for the fifth-grade class bazaar. She has a roll of 6 feet of ribbon. Each bookmark requires 4 3/4 inches of ribbon. How many (complete) bookmarks can she make? How much ribbon will she have left over?

34. Without computing common denominators, order these fractions from the smallest to the largest. Use estimation strategies, and consider the relative sizes of the numerators and denominators.

   a. 11/16, 11/13, 11/22
   b. 23/16, 33/16, 3
   c. −1/5, −19/36, −17/30
   d. 5/6, 3/4, 7/8
   e. −1/6, −1/8, 1/7
   f. 2/5, 3/10, 4/10
   g. 0/7, 0/17, 3/17
   h. −2/3, 3/4, −4/7, 4/5

35. Suppose a/b and c/d are two fractions and $a/b < c/d$. Answer each of the following questions about a/b and c/d.

   a. If $b = d$, what do you know about a and c?
   b. If $a = c$, what do you know about b and d?
   c. If $a = c$ and $b = d$, what do you know about the relationship among a, b, c, and d?

36. If the length of rod B is 2/3 unit, what is the length of rod A? Justify your answer.

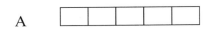

37. In a certain machine industry, a marker is a worker who draws lines on a metal blank. The blank is cut along the lines to produce the desired shape. A marker was asked to distribute seven equal-sized sheets of metal among 12 workers. Each worker was to get the same amount of metal. The marker could not use the simple solution of dividing each sheet into 12 equal parts, for this would result in too many tiny pieces. What was he to do? He thought for a while and found a more convenient method. He wrote 7/12 as 1/3 + 1/4. Then, he cut three of the sheets into fourths and four of the sheets into thirds and gave each worker one-third of a sheet and one-fourth of a sheet. How could he use his method to divide five sheets for six workers, 13 for 12, 13 for 36, and 26 for 21?

38. Suppose that $a > 1, 0 < b < 1$, and $0 < c < 2$. Fill in each ☐ with $<$, $=$, $>$ or CT (can't tell).

   a. $a \cdot b$ ☐ a b. $b \cdot c$ ☐ a c. $a \cdot b \cdot c$ ☐ b

   d. $a \div b$ ☐ a e. $a \div c$ ☐ a f. $a \div c$ ☐ b

   g. $c \div c$ ☐ c h. b^2 ☐ b

39. Ms. Jones gave her students this interesting problem: "Place a fraction in each of the squares to make each equation true. However, you may not use a fraction equal to 2 in any of the squares."

   a. ☐ · ☐ · ☐ = 2

   b. ☐ · ☐ ÷ ☐ = 2

   c. ☐ ÷ ☐ · ☐ = 2

   Jane said that the task was impossible. Mark said he could figure out the right fractions for each box! Hannah said she found several different solutions for each equation. Ms. Jones just smiled and said nothing. Who do you think is correct? Defend your choice with examples.

40. Decide whether each of the following statements is SOMETIMES, ALWAYS, or NEVER TRUE;

   a. If $x \neq 0, y \neq 0$ and $\frac{1}{x} < \frac{1}{y}$, then $x > y$.

   b. If $x > 0$, then $\frac{1}{x} < x$.

   c. If $m, n,$ and p are whole numbers, $\frac{m+n}{p+n} = \frac{m}{p}$.

41. Explain, using an illustration, why $2/3 \div 2/5 = 2/3 \cdot 5/2$

Extending the Activity

42. Use the idea of making trains with Cuisenaire Rods, and apply it to prime and composite numbers. *Hint*: Look at the number of different single-color trains that can be made for each rod (switching the order of rods does not make a different train for this activity). For example, the train shown in Activity 6.7 has only two single-color trains—the train formed by the light-green rod and the train formed by the three white rods. Write an explanation of how these trains can be used to illustrate prime and composite numbers.

Writing/Discussing

43. Make a concept map about fractions, and explain the links and connections in the map.

44. Explain why the generalization from Activity 6.7 is valid. In other words, why does increasing the length of the rod by 1 cm double the number of trains?

45. Compare and contrast operations with integers and operations with rational numbers.

46. Explain why when you divide fractions you can invert the second fraction and multiply.

47. Is it important to learn about rational numbers through illustrations? Why? How has illustrating operations with rational numbers changed your understanding of rational numbers?

CHAPTER SEVEN

Real Numbers: Rationals & Irrationals

CHAPTER OVERVIEW

As technology (especially calculators) becomes more and more prevalent for everyday use, as well as for scientific purposes, so does the use of decimals (and percent). At the same time, ratio and proportion are also quite useful to solve a variety of real-world problems. In this chapter, you will explore problems involving decimal, percent, ratio, and proportion. Furthermore, to deepen your understanding of rational numbers (which is what the foregoing kinds of numbers are!), you will study the properties of rational numbers and be introduced to still another kind of number, the irrational numbers.

BIG MATHEMATICAL IDEAS

Problem-solving strategies, conjecturing, verifying, decomposing, generalizing, representation, using language and symbolism, limit, mathematical structure

NCTM PRINCIPLES & STANDARDS LINKS

Number and Operation; Problem Solving; Reasoning; Communication; Connections; Representation

Activity **7.1** Exploring Ratio and Proportion Ideas
7.2 Solving Problems Using Proportions
7.3 Graphing Proportion Problems
7.4 Introducing Decimal Representation
7.5 Explaining Decimal-Point Placement
7.6 Modeling Operations with Decimals
7.7 Converting Decimals to Fractions
7.8 Introducing Percent Representation
7.9 Pay Those Taxes: A Game of Percents and Primes
7.10 Comparing Fractions, Decimals, and Percents
7.11 Four in a Row: A Game of Decimals and Factors
7.12 Exploring Circles: Approximating an Irrational Number
7.13 Constructing Irrational Numbers
7.14 Properties of Rational and Irrational Numbers

ACTIVITY 7.1 Exploring Ratio and Proportion Ideas

FYI Topics in the Student Resource Handbook

7.1 **Ratio**
7.2 **Proportion**

1. In a classroom, there are 15 women and 10 men.

   a. What is the ratio of women to men?

   b. Write another ratio that is equivalent to this one.

   c. What is the ratio of men to women?

   d. Write another ratio that is equivalent to this one.

   We can say that the ratios in a. and b. and in c. and d. form a **proportion.**

2. The ratio of men to women in a lecture hall is 12:5. If there are 36 men, how many women are there in the lecture hall?

   How did you solve this problem?

3. Discuss in your group and come up with a definition of the term *proportion*.

ACTIVITY 7.2 Solving Problems Using Proportions

FYI Topics in the Student Resource Handbook

7.1 Ratio
7.2 Proportion

1. During the softball season, Joan scores an average of two runs for every three runs scored by Amy. Joan scores 20 runs. How many runs does Amy score?

2. Charles is constructing a model of an early American steam locomotive, using a scale of 1:24. If the length of the original locomotive is 12 meters, what is the length, in centimeters, of the model?

3. Triangles ABC and DEF are similar. The ratio of a given side of triangle ABC to the corresponding side in triangle DEF is 4/3. If the length AB is 8 cm, what is DE?

4. The point (3, 8) lies on a straight line that passes through the origin (0, 0) on a Cartesian grid. If the x-coordinate of another point on the line is 12, what is the y-coordinate of the second point?

5. The taxes on a piece of property valued at $40,000 are $800. At the same rate, what would the taxes be on a second piece of property valued at $25,000?

ACTIVITY 7.3 Graphing Proportion Problems

FYI Topics in the Student Resource Handbook

7.2 **Proportion**

1. Graph the relationship between the scoring averages of Joan and Amy from #1 in Activity 7.2.

2. Write an equation that represents this graph. Explain what your equation means.

3. Choose two of the remaining proportion problems in Activity 7.2, and graph the relationships in the ratios. Write the equations that represent the graphs, and explain what your equations mean.

ACTIVITY 7.4 Introducing Decimal Representation

 FYI Topics in the Student Resource Handbook

 7.4 **Decimals and Fractions**

1. Write any rational number both as a fraction and as a decimal.

2. How can you prove these are the same number?

3. Model 2/5 both with a strip of paper and with base-ten blocks. Draw a picture illustrating what you have modeled.

4. Use base-ten blocks to model these operations. Draw pictures of the steps involved in performing these operations.

   a. 2.346 + 1.27

   b. 3.3 − 2.875

   c. 2 × 2.35

   d. 4.84 ÷ 4

ACTIVITY 7.5 Explaining Decimal-Point Placement

FYI Topics in the Student Resource Handbook

7.4 Decimals and Fractions
7.5 Terminating and Repeating Decimals

1. The rule for decimal-point placement in sums and differences is as follows:

   Line up the decimal points of the numbers being added or subtracted. Add or subtract as integers. Place the decimal point for the sum or difference in line with other decimal points.

   Why is this rule valid?

2. The rule for decimal-point placement in products is as follows:

   Multiply as with integers. Count the number of decimal places in both factors. Place the decimal point in the product by using the total number of decimal places in the two factors.

   Why is this rule valid?

3. The rule for decimal-point placement in quotients is as follows:

   Multiply the divisor by a power of 10 to make it an integer. Multiply the dividend by the same number. Place the decimal point in the quotient directly above the decimal point in the dividend. Divide as with integers.

   Why is this rule valid?

ACTIVITY 7.6 Modeling Operations with Decimals

FYI Topics in the Student Resource Handbook
- 7.6 **Decimal Addition and Subtraction**
- 7.7 **Decimal Multiplication**
- 7.8 **Decimal Division**

Model these products and quotients with base-ten blocks, and then draw an illustration of the steps you used to model them. For each situation, specify which block represents 1.

1. $9.6 \div 3$

2. $0.2 \cdot \dfrac{1}{3}$

3. $0.5 \cdot 0.5$

4. $0.6 \div 0.2$

5. $0.7 \div 0.3$

6. $0.2 \div 0.6$

7. $\dfrac{2}{3} \cdot \dfrac{1}{4}$

8. $1.25 \cdot 0.4$

ACTIVITY 7.7 Converting Decimals to Fractions

FYI Topics in the Student Resource Handbook

7.5 Terminating and Repeating Decimals

$$\frac{1}{2} \quad \frac{1}{3} \quad \frac{1}{4} \quad \frac{1}{5} \quad \frac{1}{6} \quad \frac{1}{7} \quad \frac{1}{8} \quad \frac{1}{9} \quad \frac{1}{10}$$

1. Change each of the fractions shown above to decimal representations.

2. In what ways can you categorize these decimals?

3. What is (are) the distinction(s) among the different categories?

4. What determines (besides dividing it out) which category a fraction will be in when expressed as a decimal? Justify your reasoning.

5. Convert these terminating decimals to fractions.
   a. 0.345

   b. 4.12

   c. −1.2359

6. Describe, in general, how to convert a terminating decimal to a fraction. Why does this method work?

7. Convert these repeating decimals to fractions. [*Hint*: Try to isolate the block of repeating digits so that you can convert this decimal to a fraction as you did above.]

   a. $0.\overline{3}$

   b. $4.\overline{45}$

   c. $1.0\overline{32}$

8. Describe, in general, how to convert a repeating decimal to a fraction. Why does this method work?

9. Can you use the same method to convert these decimals to fractions?

   a. 0.121121112...

   b. 3.515115111...

10. How can you categorize decimals of the type from #9?

11. What is the 100th digit after the decimal point in $\frac{1}{22}$?

12. What is the 200th digit after the decimal point in $\frac{1}{7}$?

13. What is the 50th digit after the decimal point in 0.121121112...?

14. What strategies did you use to find these digits?

ACTIVITY 7.8 Introducing Percent Representation

FYI Topics in the Student Resource Handbook
7.9 Percent
7.10 Interest

1. You go shopping with $60. You spend 1/4 on clothes, $30 on equipment for your home computer, and 10% of your original money on some food. How much do you have left?

2. A women bets $24 and gets back her original bet plus $48 more. She spends 25% of her winnings at a restaurant to celebrate, and 50% of her winnings to buy a present for her husband. Her paycheck was for $240, from which she made the original bet. How much money does she have left when she finally arrives home?

3. A man collects antique snuff boxes. He bought two but found himself short of money and had to sell them quickly. He sold them for $600 each. On one he made 20%, and on the other he lost 20%. Did he make or lose money on the whole deal? How much?

4. A collection of marbles has been divided into three different sets. The middle-sized set is two times the size of the smallest set, and the largest set is three times as large as the middle-sized set. What percent describes each part of the total marble collection?

ACTIVITY 7.9 Pay Those Taxes: A Game of Percents and Primes

FYI Topics in the Student Resource Handbook

7.9 **Percent**
7.10 **Interest**

Read all of the directions before starting to play.

The Game

Take turns throwing the two dice. Use the two numbers you throw to form a two-digit number that represents the amount of money you have won this turn. For example, if you throw a 2 and a 4, you may choose to win $24 or $42. However, you must pay taxes according to the following schedule.

Tax Schedule

$0–$30	pay 15% of your winnings
$31–$50	pay 25% of your winnings
$51–up	pay 35% of your winnings

Calculate your tax, and deduct this from your winnings. Round off to the nearest dollar.

Primes

After the first round of play, you can buy a prime number whenever you have enough money. The prime numbers you can buy are 2, 3, 5, 7, or 11. The prime numbers cost $500 times their reciprocal. During the game, if a player chooses to use a number (obtained from the dice) that is divisible by a prime that is owned by you, he or she must pay a commission of 50% of the winnings after taxes. If you need cash, you can sell your prime number back for its cost minus 10% interest.

Bonus

If you throw a prime number greater than 20, you get a 20% bonus before taxes. But you must pay taxes on the new number.

Winner

The winner is the player with the most money when the game is stopped, after all players sell back their prime numbers.

Record Keeping

All players must keep a neat recording of their accounts. Anyone may check another player's accounts before the turn passes and charge a $10 fee for correcting any errors found.

ACTIVITY 7.10 Comparing Fractions, Decimals, and Percents

FYI Topics in the Student Resource Handbook
6.1 **Fractions**
7.4 **Decimals and Fractions**
7.9 **Percent**

In each row, circle the item that does not belong. Justify your answer.

	A	B	C	D
1.	$\frac{3}{4}$	0.75	0.34	75%
2.	20% of 40	40% of 20	10% of 80	5% of 20
3.	1 cent	1%	1 centimeter	1°F
4.	$3\frac{1}{4}$	0.325	3.25	325%
5.	50% of 18	18 (0.5)	$\frac{1}{2}$% of 18	18% of 50
6.	$100 at 6% for 2 years	$100 at 8% for 1 year	$100 at 2% for 4 years	$100 at 4% for 2 years
7.	percent change from 80 to 120	percent change from 66 to 99	percent change from 40 to 80	percent change from 48 to 72
8.	percent change from 100 to 90	percent change from 50 to 45	percent change from 200 to 180	percent change from 60 to 50

9. List four ratios such that three are equivalent to 10% and one is not.

ACTIVITY 7.11 Four in a Row: A Game of Decimals and Factors

FYI Topics in the Student Resource Handbook
6.1 Fractions
7.4 Decimals and Fractions

Split your group into two teams, each with a calculator. In turn, each team chooses two factors (without using a calculator), one from the circular factor board below and one from the rectangular factor board. The other team multiplies the two factors to find the product. The cell on the grid that contains the product is captured by the team that chose the two factors. The first team to capture four cells in a row (vertically, horizontally, or diagonally) is the winning team. Play several games, using different group members' pages.

Grid

221.4	88.2	82.8	110.7
107.8	9	60.27	135.3
2.45	176.4	41.4	48.02
4.5	50.6	5.5	22.54

Factor Boards

Circular board: 0.9, 1.1, 0.49, 1.8

Rectangular board: 98, 46, 5, 123

What mathematical ideas are used in this game?

ACTIVITY 7.12 Exploring Circles: Approximating an Irrational Number

FYI Topics in the Student Resource Handbook

7.17 **Irrational Numbers**

1. Draw a circle, using a compass, with radius equal to the length of the red Cuisenaire rod.

    a. What is the diameter of this circle in centimeters? [The length of the white rod is 1 cm.]

    b. Find the circumference of the circle without using a formula. How did you find this?

2. In the table below, fill in the circumference and diameter information for the red rod, and calculate the sum, the difference, the product, and the quotient of C and D.

Rod	Circumference (C)	Diameter (D)	C + D	C − D	C · D	C ÷ D
red						
light green						
purple						
yellow						
dark green						
black						

3. As the size of the rod (and the circle) increases, predict whether the numbers in each column increase, decrease, or remain the same.

    C   D

    C + D   C − D

    C · D   C ÷ D

4. Draw the other circles using the rods listed in the table. Then, determine the circumference and the diameter of each of the circles in centimeters, and record in the table above.

5. Calculate C + D, C − D, C · D, and C ÷ D for each of your circles, and record these in the table above.

6. Did your predictions hold? Why or why not?

7. What number is being approximated by C ÷ D?

8. Why is this an approximation?

9. What could you do to get a better approximation?

ACTIVITY 7.13 Constructing Irrational Numbers

FYI Topics in the Student Resource Handbook

7.17 **Irrational Numbers**

1. Construct a segment that is $\sqrt{2}$ inches long.

2. Construct a segment that is $\sqrt{13}$ inches long.

3. Construct a segment that is $\sqrt{7}$ inches long.

4. Explain the strategies you used to draw these segments.

ACTIVITY 7.14 Properties of Rational and Irrational Numbers

FYI Topics in the Student Resource Handbook

7.12 **Looking at Rational Numbers as the Set of Rational Numbers**
7.18 **Real Numbers**

1. Justify whether or not the following properties are valid for the set of rational numbers.

   a. Closed under addition

   b. Closed under multiplication

   c. Commutative Property for addition

   d. Commutative Property for multiplication

   e. Associative Property for addition

   f. Associative Property for multiplication

   g. Identity element for addition

   h. Identity element for multiplication

   i. Inverse elements for addition or multiplication

2. Justify whether or not these properties are valid for the set of irrational numbers.

3. The set of real numbers is the set of all rational numbers and all irrational numbers joined together. Justify whether or not the above properties are valid for the set of real numbers.

4. Use a diagram of some sort to illustrate the relationships among all the sets of numbers that you know.

Things to Know from Chapter 7

Words to Know

- associativity
- closure
- commutativity
- decimal (terminating; repeating; nonterminating, nonrepeating)
- identity
- inverse
- irrational number
- percent
- pi (π)
- proportion
- ratio
- rational number
- real number

Concepts to Know

- what is a ratio
- what is a proportion
- what is a decimal
- why are decimal point placement rules valid
- why can some fractions be represented as terminating decimals while others cannot
- what is a percent
- what is a rational number
- what is an irrational number
- which group properties are valid for rational numbers, irrational numbers, real numbers

Procedures to Know

- representing quantities as ratios
- solving proportions
- representing numbers in decimal notation
- converting fractions to decimals and decimals to fractions
- representing quantities as percents
- performing operations with decimals and percents

Exercises & More Problems

Exercises

1. Illustrate what a base-ten block model of 1/5 would look like.

2. Compute each of the following:

    a. $4.243 + 1.01$
    b. $1.32 - 2.1$
    c. $-4.208 - 1.002$

3. Compute each of the following:

    a. $4 \cdot 3.021$
    b. $5.2 \div 3.2$
    c. $-12.24 \div 2.04$

4. Find a rational number, x, such that $7/36 < x < 5/24$. Justify your answer.

5. Arrange these decimals in order from the smallest to the largest.
   a. $3.2, 3.\overline{22}, 3.\overline{23}, 3.2\overline{3}, 3.23$
   b. $-1.454, -1.45\overline{4}, -1.45, -1.4\overline{54}, -1.\overline{454}$

6. Arrange these decimals in order from the largest to the smallest.
   a. $0.002, 0.\overline{02}, 0.02\overline{5}, 0.0\overline{02}, 0.02$
   b. $-1.\overline{19}, -1.2\overline{1}, -1.19, -1.192, -1.\overline{21}$

7. What strategies did you use to determine the order of several decimals?

8. Find a number between 0.84 and 0.85

9. Find a number between -2.3 and -2.29

10. Place the following decimals in order from the least to the greatest:
    a. $3.29, 3.231, 3.2957, 3.23057, 3.230525$
    b. $-0.7\overline{3}, -0.737, -0.7373, -7.\overline{37}$

11. Which of the following are true:
    a. $7/8 > 6/7 > 15/28 > 6/10$
    b. $-1.88 < -1.871 < -1.8701 < -1.8$
    c. $3.2 < 3.\overline{22} < 3.23 < 3.\overline{23}$
    d. a and b
    e. b and c
    f. a, b, and c

12. Base-ten blocks include units, longs, flats, and cubes (1,000 units).
    a. If 7 longs are used to represent the decimal 0.7, which block is equal to 1?
    b. Using the cube as 1, represent the following sum and its solution: $2.34 - 0.53$.

13. Write the following fractions as terminating or repeating decimals:
    a. 7/8   b. 1/3   c. 11/12

14. Write the following terminating decimals as simplified fractions:
    a. -5.32 b. 4.022   c. 25.15

15. Write the following repeating decimals as fractions.
    a. $0.4\overline{3}$ b. $0.\overline{35}$

16. Determine which of the following represent irrational numbers:
    a. $\sqrt{29}$ b. $\sqrt{121}$ c. $(\sqrt{3})/2$

17. Determine two irrational numbers that exist between 3/4 and 1.

18. Is $\dfrac{\sqrt{48}}{\sqrt{108}}$ a rational number? Explain.

19. Draw a Venn diagram to indicate the relationships among the set of whole numbers, integers, rational numbers, natural numbers, irrational numbers, and real numbers.

Critical Thinking

20. Put numbers in the boxes that will give you the desired quotient.

    ☐ ÷ ☐ has a quotient between 3 and 4.

    ☐ ÷ ☐ has a quotient between 0 and 1.

    ☐ ÷ ☐ has a quotient between 8 and 9.

21. Number suspects: 1/2, 3/4, 2/5, 3/5, 5/8, 7/10, 12/11.

    Clue 1: I am greater than 0.5.
    Clue 2: I am not equal to 0.75.
    Clue 3: If you multiply me by 2 you get a number that is less than 2.
    Clue 4: My denominator is a prime number.
    Who am I? ____

22. Make up a Number Mystery similar in structure to the one in #21. Name your own number suspects, and write at least 4 clues.

23. If the ratio of boys to girls in a class is 3:8, will the ratio of boys to girls remain the same, become greater, or become smaller if 2 boys and 2 girls leave the class. Justify your answer.

24. How do you know that a number is irrational (what classifies it as such)?

25. How do repeating decimals affect the ordering of decimal values?

26. Explain the procedure used to convert repeating decimals to fractions.

27. When multiplying and dividing decimals, why is it not important to line up the decimal points as when adding or subtracting?

28. Using some of the numbers 0.1, 0.2, 0.3, 0.4, 0.5, 0.6, 0.7, 0.8, and 0.9 and one or more of the four operations, represent each of the following:

    a. 1
    b. a number between 0.5 and 1
    c. a number between 0.4 and 0.5
    d. a number greater than 3
    e. a number between 0.1 and 0.2
    f. 0
    g. 1.5
    h. a number between −3 and −5

29. The answers to the computations below are estimates in which the decimal points have been left out. Place the decimal point where it belongs in each estimate. Justify your placement of the decimal point.

    a. $5.03 \times 17.6 = 886$
    b. $68.64 \div 4.4 = 156$
    c. $6.23 \times 17.91 \times 0.131 = 146$
    d. $16.55 \times 0.008 = 132$
    e. $0.726 \div 0.154 = 471$
    f. $400.14 \div 85.5 = 468$
    g. $0.198 \div 0.090 = 22$
    h. $0.7265 \div 0.954 = 7615$
    i. $5.824 \times 0.36 = 209664$

30. Candy bars priced 50¢ each were not selling, so the price was reduced. Then they all sold in one day for a total of $31.93. What was the reduced price for each candy bar? Explain.

31. Explain why 3/7 cannot be written as a terminating decimal.

32. Rebekah ate five pieces of pizza, and Moses ate four pieces. Rebekah says she ate 25% more than Moses, but Moses says he ate only 20% less than Rebekah. Who is correct? Justify your answer.

33. A collection of marbles has been divided into three different sets. The middle-sized set is two times the size of the smallest set, and the largest set is three times as large as the middle-sized set. What percent describes each part of the total marble collection?

34. Is it possible to have a square with area $5\,\text{cm}^2$? Justify your answer.

35. Is it possible to have a square with area $11\,\text{cm}^2$? Justify your answer.

36. How could triangles help to represent irrational numbers (more specifically, square roots)?

37. If a 10×10 square of cubes is built (only 1 block high), glued together, and suspended in the air and then the entire group of blocks is painted, what percent of the cubes will have the following?

    a. four faces painted b. three faces painted c. two faces painted

38. Answer the question in #37 for squares of cubes of the following dimensions.

    a. 9×9 b. 8×8 c. 7×7
    d. 12×12 e. $n \times n$

Extending the Activity

39. Give three different examples that prevent you from thinking that all numbers are rational.

40. Which digit between the 50th and 60th digits after the decimal point in 1/17 is 9? Justify your answer.

Writing/Discussing

41. If the ratio of green marbles to red marbles in a collection is 4:7, explain how this ratio is different than saying that the green marbles are 4/7 of the collection.

42. Explain the procedure for changing a fraction to a decimal.

43. Explain why not all fractions can be changed to terminating decimals. Include in your explanation a discussion of how you can determine if a fraction can be changed into a terminating decimal.

44. Explain the procedure for changing a terminating decimal to a fraction.

45. Explain the procedure for changing a repeating decimal to a fraction.

46. Do you think it is possible, for every irrational number, to construct a length equal to that number in some unit such as inches? Why or why not?

47. Explain the relationships among fractions, decimals, and percents.

48. Make a second concept map about fractions, and explain the links and connections in the map. Then write a reflection comparing and contrasting your first and second concept maps.

CHAPTER EIGHT

Patterns & Functions

CHAPTER OVERVIEW

The concept of function is a central theme, a big idea, running through many areas of mathematics. In this chapter, the idea of function, which is a particular kind of relationship between two or more sets of objects, is introduced through explorations of patterns. Also emphasized will be different ways to represent functional relationships.

BIG MATHEMATICAL IDEAS

Problem-solving strategies, functions and relations, representation, conjecturing, verifying, mathematical structure

NCTM PRINCIPLES & STANDARDS LINKS

Patterns, Functions, and Algebra; Problem Solving; Reasoning; Communication; Connections; Representation

Activity **8.1** Exploring Variables
8.2 Investigating Variables Through Data in Tables
8.3 Investigating Variables Through Data in Graphs
8.4 Interpreting Graphs
8.5 Investigating and Describing Numerical Patterns
8.6 Investigating Numerical Situations
8.7 Identifying Rules and Functions
8.8 Looking for an Optimal Solution
8.9 Investigating Numerical Functions That Repeat
8.10 Investigating Real-Life Iteration: Savings-Account Interest
8.11 Investigating Iteration: Geometry and Fractals
8.12 Properties of Equations
8.13 Using Properties to Solve Equations
8.14 Investigating Distance vs. Time Motion
8.15 Constructing and Interpreting Sensible Graphs

ACTIVITY 8.1 Exploring Variables

FYI Topics in the Student Resource Handbook
8.1 Functions

The temperature on a cold winter day in Syracuse, New York, is measured for a 24-hr period. The table below gives the readings (time, temperature) for this winter day from 1:00 A.M. to midnight.

Time (hours)	Temperature (°F)	Time (hours)	Temperature (°F)	Time (hours)	Temperature (°F)
1	14	9	17	17	12
2	12	10	20	18	8
3	10	11	24	19	5
4	8	12	29	20	2
5	8	13	32	21	0
6	9	14	31	22	−1
7	10	15	27	23	−1
8	13	16	18	24	1

Use the data in the table to answer the following questions.

1. a. What was the temperature at noon?

   b. At which time(s) was the temperature 13°F?

2. a. How much does the temperature decrease between 2 P.M. and 6 P.M.? How does this change compare to the decrease between 2 P.M. and 4 P.M.? Explain.

b. What is the increase between 6 A.M. and noon?

3. During which time period was the temperature dropping fastest? Explain.

4. Use the grid below to plot the data pairs (time, temperature) using the data in the table.

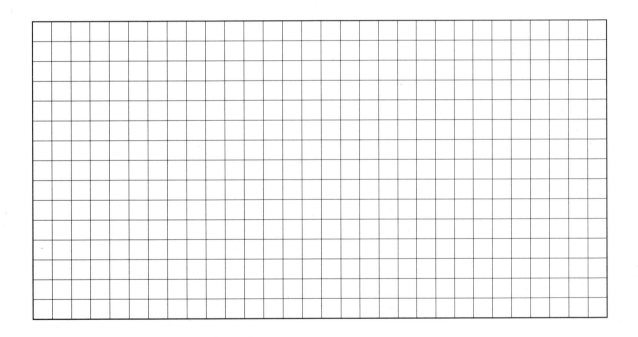

5. a. At what time was the temperature the highest?

b. At what time was the temperature the lowest?

6. For each part a through d:
   i. Write the value that correctly completes the equation.
   ii. Give the coordinate of the data point (input, output) that supplies your answer.
   iii. Write the completed statement in words.

   a. $T(7) = \underline{}$

   b. $T(19) = \underline{}$

   c. $T(\underline{}) = 0$

   d. $T(\underline{}) = 29$

7. Discuss the factors that affected the temperature during the day. Explain the temperature pattern of the day.

ACTIVITY 8.2 Investigating Variables Through Data in Tables

FYI Topics in the Student Resource Handbook

8.1 **Functions**
8.2 **Equations and Inequalities**

A movie theater finds that by changing the price on a small bag of popcorn, they will change the number of bags they sell during a movie.

Cost ($)	# of bags sold during movie
$1.90	100
$1.95	95
$2.00	90
$2.05	85
$2.10	80
$2.15	75

1. What relationships do you notice between the cost and the amount of popcorn the theater sells?

2. Explain why these relationships seems reasonable (or unreasonable).

3. Is the information in the table above sufficient to determine the price that theaters should choose for a bag of popcorn? Explain your reasoning.

4. Suppose it costs the theater $1.25 to make and sell a small bag of popcorn. Make a third column in the table above that shows the amount of *profit* the theater makes for each selling price. Based on this, what would be the best price for the theaters to choose?

Because it seems that the number of bags sold depends on the cost of a bag, we could use *function notation* to describe the data in the above table. For example, to say that "If the cost is $2.10, the theater will sell 80 bags of popcorn," we could write B($2.10) = 80, where B represents the number of bags sold at the price (given in parentheses). Note that whenever you begin to use a variable to represent something, you should let others know what that variable means. Because we have already stated what B stands for, you may continue to use it without needing to specify its meaning.

5. Use functional notation to represent each of the following sentences:

   a. When a bag of popcorn costs $1.90, 100 bags of popcorn will be sold.

   b. The theater only sells 75 bags when the cost is raised to $2.15.

6. Use functional notation to show the number of bags you think would be sold if the cost were raised to $2.40.

7. If we let P represent the *profit* earned, then write in words what $P(\$1.90) = \$65$ means.

8. Use functional notation to tell the theater the best possible cost if they wish to maximize profit.

9. What other factors besides cost could influence the number of bags of popcorn sold during a movie?

10. What are some possible consequences of making the cost of a bag of popcorn too *high*?

11. What are some possible consequences of making the cost of a bag of popcorn too *low*?

ACTIVITY 8.3 Investigating Variables Through Data in Graphs

Topics in the Student Resource Handbook
- 8.1 **Functions**
- 8.2 **Equations and Inequalities**

In the desert, temperatures can range from extremely hot during the day to extremely cold at night. Below are some sample time and temperature data from a desert.

Time (hours)	Temperature (°F)
0 (midnight)	35
2	30
4	38
6	61
8	70
10	85
12 (noon)	98
14	96
16	88
18	66
20	48
22	40

Plot the above data in a line plot, with "Time" on the horizontal axis and "Temperature" on the vertical axis. You may want to use graph paper, a spreadsheet, or a graphing calculator for your plot. Answer the questions below, stating the coordinates (when appropriate) that you used to find your answers. Coordinates should be given as ordered pairs (time, temperature).

1. Why is a line plot more appropriate than a bar graph for the given data?

2. What was the temperature at 4 P.M.?

3. At what times was the temperature in the 60s?

4. At what times was the temperature below 40°?

5. Was the temperature ever 73°? Explain your reasoning.

6. What was the change in temperature between 8 and 10 P.M.?

7. What was the *rate of change* (in degrees per hour) between 8 P.M. and 10 P.M.?

8. What might the temperature have been at 7 A.M.?

9. During which two-hour time period was the temperature *rising* most rapidly?

10. During which two hour time period was the temperature *falling* most rapidly?

11. In terms of the appearance of the line plot, explain how you can tell when the temperature is *rising* most rapidly.

12. What might cause the drop in temperature that occurred between 4 P.M. and 6 P.M.?

13. Are the data in this table sufficient for determining the highest temperature in the desert for the day this data was collected? Why or why not?

ACTIVITY 8.4 Interpreting Graphs

FYI Topics in the Student Resource Handbook

8.3 Cartesian Coordinate Systems
8.4 Graphs and Equations of Lines

1. In general, sales of a product depends on its price. The diagrams below show graphs of three possible relations between the price of a house and sales. No scales are given on the axes, but following the conventions that right and up represent positive values of price and sales respectively, you should be able to decide which graph seems most likely to match the relation between price and sales.

   a. For each diagram, explain why you think the graph does or does not fit the likely relation between price and sales.

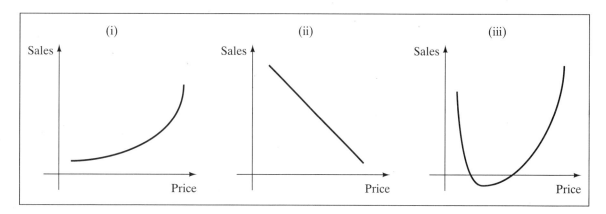

   b. List at least four factors other than price that are likely to affect the sales price of a house.

c. Pick a factor from your list that is a quantitative (numerical) variable. Sketch a graph similar to the ones above that shows the relation you would expect between that variable and house sales. Explain in words the relation shown by your graph (that is, tell what happens to house sales as the input variable increases), and give some possible reasons for this relation.

Identifying Graphs of Functions

Suppose you were to graph the height of water in a bathtub while it was filling, being used, and emptying. What would that graph look like? In each of the problems below, you are presented with a scenario and three possible graphs. For each scenario:

a. Choose the graph that you feel best represents that scenario. Be prepared to explain your selection.

b. For each of the graphs you did *not* select, write an explanation of how that graph could have come about.

1. The height of water in a bathtub during the time that it is filling, being used, and emptying:

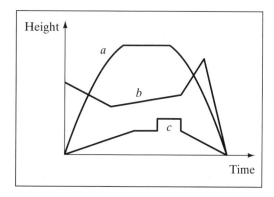

2. The daily high temperature recorded in Chicago, Illinois, from January to December.

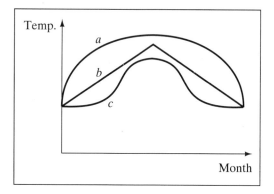

3. The fraction of the moon's surface that is visible over a two-month period.

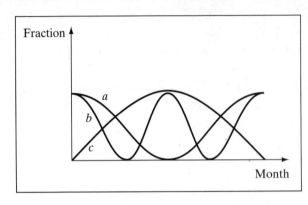

4. The number of apples you can buy with $5 as the cost of apples increases.

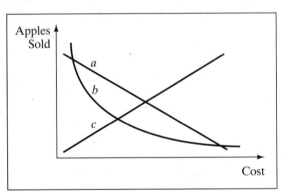

ACTIVITY 8.5 Investigating and Describing Numerical Patterns

FYI Topics in the Student Resource Handbook
8.2 Equations and Inequalities

For each input-output table below:
- Complete the table.
- Describe how the output number is related to the input number.
- Answer any questions included.

1.

Input	Output
1	3
2	6
3	9
4	12
5	
6	
⋮	⋮
	51

a. In words, describe how the output number is related to the input number.

b. If the input number were x, what would be the output number? _____

c. If the output number were n, what must be the input number? _____

2.

Input	Output
1	19
2	18
3	17
4	
10	
	6.5
0	

a. Describe the relationship.

b. If the input number were h, then the output number would be _____.

c. If the output number were j, then the input number was _____.

3.

Input	Output
1	1
6	11
20	39
13	25
5	
	41
	22

a. Describe the relationship:

b. If the input number were z, then the output number would be: _____.

c. If the output number were d, then the input number was: _____.

4.

Input	Output
3	12
7	56
1	2
0	0
25	
26	
	2256

a. Describe the relationship:

b. If the input number were x, then the output number would be: _____.

c. If the output number were y, then the input number was: _____.

5. Make up a table of your own, and describe a rule for that table. Be prepared to trade your table (but not your rule) with another group.

Input	Output

ACTIVITY 8.6 Investigating Numerical Situations

FYI Topics in the Student Resource Handbook
  8.2 **Equations and Inequalities**

Suppose you are the director of the soccer for the City Park Summer Programs. In this league, each team plays every other team once before the playoffs, so the number of games to be scheduled (not including the playoff games) depends upon how many teams are allowed in the league. Your immediate task is to plan the soccer league.

1. To get a feel for the numbers involved, do some calculations. It will be helpful for you to record your findings in a table. It may also be helpful to think of ways to draw diagrams to model the problem situation.

2. Make a graph by plotting the data points from your table. If you have access to a spreadsheet or a graphing calculator, you may want to use it to create your graph. Does it make sense to connect the points together? Why or why not?

3. What type of relationship do you think there is between the number of teams and the number of games played?

4. How can you use the table and scatter plot to obtain more data points without doing the types of calculations you did in #1?

5. If there are *n* teams in the league, find a formula for the number of games needed. Explain how you found this formula.

6. Use your formula to calculate the number of teams that can be in the league if the maximum number of games you can schedule is 120.

ACTIVITY 8.7 Identifying Rules and Functions

FYI Topics in the Student Resource Handbook
 8.2 Equations and Inequalities
 8.4 Graphs and Equations of Lines

1. You can calculate the speed of a car by knowing the size of its tires and the revolutions per minute that the rotating tires make. Suppose the tires of a car cover about 2.5 meters of ground in one revolution and the tires are rotating at 500 revolutions per minute.

    a. What is the speed of the car?

    b. Write verbal and symbolic rules expressing the relation between time and distance.

    c. Calculate some specific data pairs (time, distance), and record them in a table.

d. Write the calculation required to find the distance traveled in 20 minutes. What is this distance?

e. Sketch a graph showing the pattern of the relation between time and the distance traveled. The input (independent) variable should be represented on the horizontal axis, and the output variable (dependent) should be on the vertical axis.

2. Regina has a part-time job working for a national company selling Internet subscriptions house to house. She is paid $100 per week plus $15 for each subscription she sells.

   a. Complete the table below, showing her weekly pay as a function of the number of subscriptions she sells.

Number of subscriptions sold	Weekly pay (dollars)
1	
2	
3	
6	
10	
12	
15	

b. Using variables, write two symbolic rules, one for the pay in terms of sales and one for sales in terms of pay.

c. Write the calculation required to answer each of the following questions about Regina's pay prospects. Then, find the numerical result.

  i. How much does she earn if she sells nine subscriptions in a week?

  ii. How much does she earn if she sells no subscriptions in a week?

  iii. Regina's goal is to earn $400 dollars next week. How many subscriptions does she need to sell if she wants to accomplish this goal?

ACTIVITY 8.8 Looking for an Optimal Solution

FYI Topics in the Student Resource Handbook
8.5 Systems of Linear Equations

Below are the directions for making a box out of a sheet of paper. Follow these directions to create the box; then answer the questions that follow.

Making a Paper Box

Materials: You will need an 8-inch by 10-inch piece of paper, a ruler, scissors, and tape.

Directions:

- Begin by tracing 1-inch squares in each of the four corners of the paper, as shown:

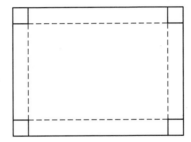

- Next, cut out these four squares from the corners, and discard them. Finally, fold and crease along the indicated dotted lines, and tape the edges to form a box (with no lid).

1. What is the volume of the box you constructed? What is the outer surface area? Explain how you arrived at your answers.

2. What if, instead of cutting out squares that measured 1 inch on each side, you cut out squares that measure 2 inches on each side. What would be the resulting volume and surface area?

3. What is the largest square that you could cut from each corner? What would be the resulting volume and surface area?

4. What size square would you have to cut from each corner to maximize the volume? Explain how you went about finding your answer.

5. What size square would you have to cut from each corner to maximize the surface area? Explain how you went about finding your answer.

6. What if you cut out a square from each corner whose sides measured x inches. What would be the resulting volume and surface area?

7. Without actually measuring, what would be a *practical* way to compare the volumes of two of these paper boxes?

Shown below is an aerial view of Farmer MacDonald's barn. She wishes to put up a rectangular pen for her goats so that one side of the pen abuts one side of her barn, as shown:

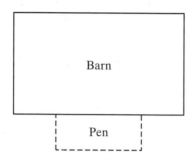

By using one side of the barn as one side of the pen, Farmer MacDonald realized that she will only need fence for the other three sides. Knowing that the long side of the barn measures 200 feet and that Farmer MacDonald has 120 feet of fencing she can use, answer the questions below.

8. In the figure shown, the left and right sides of the pen must be the same length. What is the longest each of those two sides may be?

9. What is the longest that the front side of the pen may be? Explain how you arrived at your answer.

10. If MacDonald wanted a square pen that used all of the fencing, what would be the area of the ground within the pen?

11. Creating a rectangular pen as described above, what is the maximum area MacDonald can get by using all 120 feet of fencing? Be ready to explain how you went about finding this solution. How do you know that your solution represents the *maximum* area?

12. Creating a rectangular pen as described above, what is the minimum area MacDonald can get by using all 120 feet of fencing? Be ready to explain how you went about finding this solution. How do you know that your solution represents the *minimum* area?

13. What if she decided that, instead of building the rectangular pen adjacent to the barn, she would build it in the middle of a field? Make a table that shows the relationship between the length of one side of the fence and the area of the pen. What would be the minimum and the maximum areas possible, using all 120 feet of fencing?

ACTIVITY 8.9 Investigating Numerical Functions that Repeat

FYI Topics in the Student Resource Handbook

- 8.2 Equations and Inequalities
- 8.5 Systems of Linear Equations
- 9.25 Fractiles and Rep-tiles

1. Get a calculator. Pick a number (any number), and enter it on the calculator. Square it. Now, take the result and square it. Square the new result. Square it again. And again. Describe what is happening to the size of the number.

2. Square the result a few more times. What would happen if you could do this *forever*? What would the "long-term" behavior be?

3. So you picked a number, repeatedly squared it, and described the long-term behavior. Try this again but with a different "seed" (starting number). Is the long-term behavior the same?

4. See if you can get different long-term behaviors by starting with different seeds and repeatedly squaring the results. Try to classify all real numbers according to their long-term behavior in this process.

5. Suppose someone makes the statement that whenever you square a number, the result will be a bigger number. Based on your explorations above, what could you say to that person?

6. Explore: What would happen if, instead of starting with a seed and repeatedly *squaring* the results, you instead *cubed* the results? In what ways would your results be the same as above? In what ways would the results be different?

More Iteration

This one is only a little more complicated, and the pattern may surprise you. You will need a calculator again. It may be helpful to know how to use the memory function of your calculator.

1. Pick a number.
2. Enter that number into the expression $\frac{x}{2} + \frac{8}{x}$. (*Store* the result on your calculator.)
3. Write down the result.
4. Take this result (*recall* from your calculator), and reenter it into the same expression, $\frac{x}{2} + \frac{8}{x}$.
5. Repeat, writing your result at each stage.

Here is an example, assuming you started with the number 17.

Start with	End with
17	8.970588235
8.970588235	5.377097396
5.377097396	

Look at the list of numbers we have generated. What do you notice about the numbers?

Activity 8.9 *Investigating Numerical Functions that Repeat* **319**

1. Try the same exploration again but with a different starting number. Keep track of the numbers you generate. Perform as many steps as it takes to begin to see what you might call predictable long-term behavior. What relationships and patterns do you notice about the seed (starting number) and the long-term behavior of the list of numbers you generate?

2. Try the exploration again but using the expression $\dfrac{x}{2} + \dfrac{1}{x}$. Begin with different seeds, each time generating a list of numbers. Describe the patterns and relationships you notice.

3. Notice that the long-term behavior seems to depend on the expression we use. When we changed from $\dfrac{x}{2} + \dfrac{8}{x}$ to $\dfrac{x}{2} + \dfrac{1}{x}$, the long-term behavior changed. Modify this expression again, and explore the expression you've created by trying different seeds and exploring the long-term behavior.

ACTIVITY 8.10 Investigating Real-life Iteration: Savings-Account Interest

FYI Topics in the Student Resource Handbook
- 8.6 **Solving Systems of Linear Equations Symbolically**
- 9.25 **Fractiles and Rep-tiles**

To *iterate* something means to *repeat* it, which is exactly what you did when you repeatedly squared numbers in the previous activity.

"Iteration" is a good way to describe the way a bank calculates interest. Let's take a simple example of a bank that pays you 6% interest, compounded annually, on money in your savings account. *Compounded annually* means that once a year, the bank will calculate 6% of what you have in your account (actually, the bank figures out the *average* amount that has been in your account) and adds that to your total. So if you had $100 in your account for one year, how much would you have after the bank added interest?

Now, if you don't touch your savings-account money for another year, the bank will *again* calculate 6% interest. But this time, they won't use $100 as the starting amount. Instead, they will give you interest on the $100 *plus last year's interest*. So how much will you have in your account after the second year?

Suppose, now, that you just leave your savings account alone, allowing the interest to accumulate. Determine how much money you will have in your account (remember that you started the first year with $100) after each of the first 5 years.

Year	Total After That Year
1	
2	
3	
4	
5	

1. How much will be in your account after 10 years? 20 years?

2. Generalize a way of finding the amount of money in your account after any number of years.

3. How many years would it take for your money to double?

4. What if the annual interest rate were 4%? How much would be in the account after 10 years (having started with $100)? After how many years would your money double?

5. Find a relationship between the annual interest rate and the number of years it takes for your money to double.

6. Points to Ponder:

    What does it mean if a bank compounds interest semiannually? Demonstrate that compounding interest semiannually results in more interest than compounding annually. How do banks calculate interest? What is the difference between annual percentage *rate* and annual percentage *yield*?

ACTIVITY 8.11 Investigating Iteration: Geometry and Fractals

FYI Topics in the Student Resource Handbook
9.25 **Fractiles and Rep-tiles**

Begin with a line segment (this will be your "seed").

———————————

Now, *trisect* (cut into three equal parts) the segment:

———————————

Finally, create an equilateral triangle "bump" on the middle section: then erase that middle section:

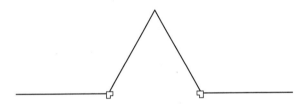

Now, with our four new, shorter segments, we can *iterate* the same two steps as above, giving us this figure:

1. Using triangular graph paper, create the above figures, as well as the next three stages. Decide in advance on a segment length that will be easy to trisect; then trisect each smaller part, and so on.

2. Suppose the length of the original segment were 1 unit (we could call this Stage 0). Find the total length of each of the figures you drew.

Stage	Total length (in "units")
0	1
1	
2	
3	
4	
5	

3. Find a way to calculate the length after any stage. After how many stages would the path length be longer than 10 units? Longer than 100 units?

Iteration and Koch's Snowflake

Using simple iterative procedures, complex geometric and numeric patterns can arise.

Using triangular grid paper, construct an equilateral triangle with sides of 27 units. Orient your triangle so that the "top" vertex is near the top center of the paper.

For this activity, you will be measuring perimeters and areas of figures you construct. For our purposes, we will measure lengths in "units" where one unit is equivalent to the length of one side of one of the small triangles. We will measure area in "triangular units" that are equivalent to the area of one of the small triangles.

1. What are the perimeter (in units) and the area (in triangular units) of the triangle you constructed?

2. Now, as we did in the previous section, trisect each segment of your figure, and construct an equilateral bump on each middle section of the segments. These bumps should point outward so that the resulting figure looks like a six-pointed star. What are the perimeter and the area of this star?

3. Repeat the above steps by trisecting each of the smaller segments, constructing the bumps, and so on. What are the new perimeter and area?

4. Repeat again, making note of the perimeter and the area.

5. Complete the table below, listing the perimeters and areas for each stage so far, starting with stage 1 (the big triangle).

Stage	Perimeter (Units)	Area (Triangular Units)
1		
2		
3		
4		
5		
6		

6. Describe the "behavior" of the perimeter and the area as you progress through the stages. Do you have enough information to make any long-range predictions? Do the perimeter and the area "behave" in similar ways?

7. Someone makes the claim that, even though the area is growing larger, it can't go beyond a certain value. How would you respond to this claim? Could a similar claim be made about the perimeter?

ACTIVITY 8.12 Properties of Equations

FYI Topics in the Student Resource Handbook

8.5 Systems of Linear Equations

Addition Property of Equations

Solve the following equations for the variable.

1. $x + 3 = 93$
2. $a - 8 = 10$

3. $3w - 14 = 1$
4. $-2b + .5 = 2.5$

To find the values of the variables in the equations above, you used the Addition Property of Equations. Using what you know about addition and equations, make a conjecture about what the Addition Property of Equations is. Try to state it as precisely and concisely as possible.

Multiplication Property of Equations

To solve #3 and #4 above, you used the Multiplication Property of Equations. Using what you know about multiplication and equations, make a conjecture about what the Multiplication Property of Equations is. Try to state it as precisely and concisely as possible.

Substitution Property of Equations

The following two examples make use of the Substitution Property of Equations.

5. $2(x + 4) = 20$ and $2x + 8 = 20$ 6. $(b + 3)(b - 3) = 0$ and $b^2 - 9 = 0$

Using what you know about substitution and equations, make a conjecture about what the Substitution Property of Equations is. Try to state it as precisely and concisely as possible.

ACTIVITY 8.13 Using Properties to Solve Equations

FYI Topics in the Student Resource Handbook

8.5 Systems of Linear Equations
8.6 Solving Systems of Linear Equations Symbolically

Solve each equation for the variable by showing all the steps involved. State the property you are using at each step.

1. $3x + 5 = 2x - 8$

2. $-0.5a + 36 + 4a = -9a - 12$

3. $5(c + 3) - 2c + 9 = 2(c - 1) + 13$

ACTIVITY 8.14 Investigating Distance vs. Time Motion (Optional—Requires TI-82, 83, 89, or 92 calculator and CBL or CBR Unit)

FYI Topics in the Student Resource Handbook

8.1 **Functions**

Experiment Setup Procedure

1. Instructor downloads program HIKER into one calculator in each group. Collected data will be stored on this calculator. After data collection is complete, each group will download data and program so each calculator has the same data.
2. Connect the CBL unit to the calculator with the unit-to-unit link cable using the I/O ports located on the bottom edge of each unit. Press the cable ends in firmly.
3. Connect the motion detector to the SONIC port on the left side of the CBL unit.
4. Turn on the CBL unit and the calculator.
5. Place the motion detector on the table so that it will detect the movement of a student walking away from or toward the detector.

The CBL system is now ready to receive commands from the calculator.

Experimental Procedure

1. Clear a walkway in the area designated by your instructor. Make sure that the walkway is wide enough so that no one else except the walking student will have his or her motion detected by the motion detector.
2. The motion detector should be perpendicular to the line of motion of the walker and sitting flat on the tabletop.
3. The walking student must stay in the motion detector's beam and walk perpendicular to the detector. **Note:** The student must walk beyond 1.5 feet from the motion detector because the detector cannot detect objects closer than that distance.
4. The detector takes measurements in feet every 0.1 seconds for 6 seconds and displays the graph of the collected data on the calculator.
5. Make sure the CBL is turned on. Start the program HIKER on the TI-82. The program will pause with a message "PRESS ENTER TO START GRAPH". When you and the walker are ready, tell the walker to start to walk after you press ENTER and hear the clicking sound from the motion detector.
6. A graph of Distance (y-axis) vs. Time (x-axis), is plotted on the TI-82 as the data is collected. When the walker has a reasonably good graph, move to the analysis section of the lab (below). To obtain the data (ordered pairs) needed for the analysis, press TRACE, and move along the plotted points.

Analysis

1. Take turns being the walker for each of the following graphs. Draw a rough sketch of your graph, and discuss with your group what the walker had to do to obtain the graph. Write a careful and complete explanation of what the walker did.

   Express the Distance as a function of Time either by using TRACE on the graph to find the coordinates of endpoints of the line or by analyzing the data table if you have stored it already.

a. Line with a positive slope

b. Line with a steeper positive slope

c. Line with a negative slope

d. Line with a less-steep negative slope

e. Line with a slope of zero

f. Parabola opening down

2. Compare and analyze all the graphs you made, and answer the following questions:

   a. What does a graph look like when a person is not moving? Give some examples.

   b. How can you tell from a graph whether a person is walking or running? Give some examples.

   c. How can you tell in which direction a person is moving from a graph?

d. How can you tell where a person started from by looking at a graph?

3. The following graphs were obtained using the CBL-motion detector system (distance vs. time). Describe as fully as possible the walks represented by these graphs, and explain the differences between them.

   a.

   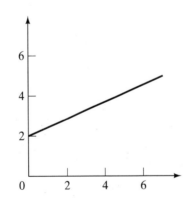

b.

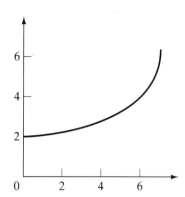

c.

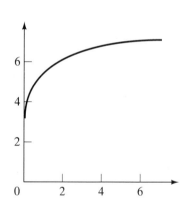

ACTIVITY 8.15 Constructing and Interpreting Sensible Graphs

FYI Topics in the Student Resource Handbook

8.3 **Cartesian Coordinate System**
8.4 **Graphs and Equations of Lines**

Part A: Interpreting graphs. It is one thing to be able to construct a graph of a relationship and another to be able to interpret the information presented in a graph. In this activity, you are given five statements involving a relationship between elapsed time and another variable. For each statement, you are to identify the graph among a set of four graphs that best represents the relationship implied.

1. A bus drives into the bus station and drops off its passengers.

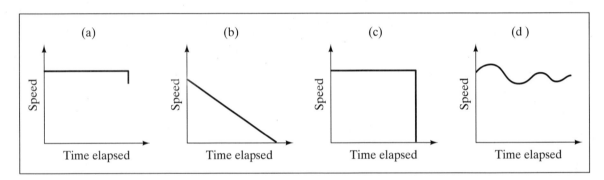

Justify your choice of graph.

2. A baby climbs up to the top of a sliding board and then slides down it.

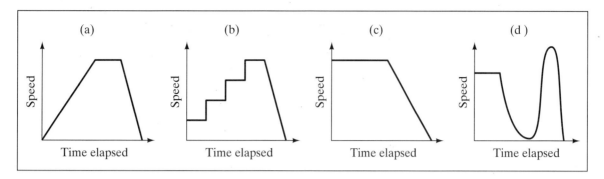

Justify your choice of graph.

3. A boy walks up a hill at a constant rate and then runs down the other side.

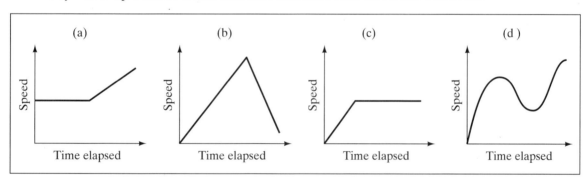

Justify your choice of graph.

4. A girl takes a ride on a Ferris wheel.

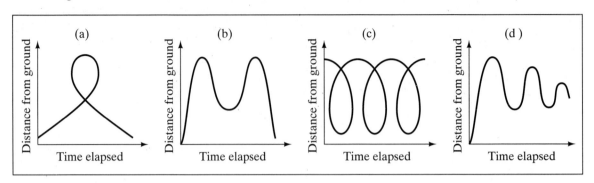

Justify your choice of graph.

5. A small child swings on a swing in the park.

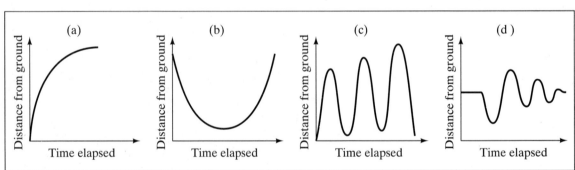

Justify your choice of graph.

Part B. Constructing graphs of everyday situations. Think about what is going on in the situation described in the statement your group has been given. After you have thought about the situation, construct a graph, using the axes shown on the next page to represent the number of people present in the situation throughout a 24-hour period. *[Note: You should NOT label the vertical (number of people) axis with your situation.]*

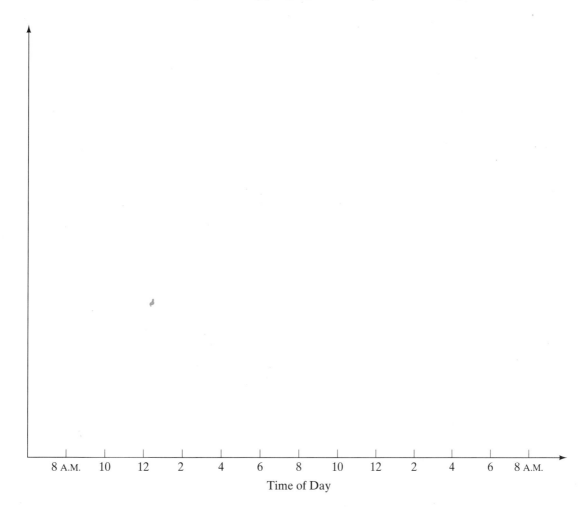

1. Prepare to explain why your graph looks the way it does. (Pay special attention to points on your graph where there are dramatic increases or decreases in number of people.)

2. Study the graphs of each of the other groups in the class. Prepare a set of questions to ask the other groups about the their graphs.

3. Identify the situations depicted in the graphs of the other groups in the class. Justify why you think each graph represents the situation you think it does.

4. Why do you think you were told not to label the vertical axis of your graph?

Things to Know from Chapter 8

Words to Know

- data
- dependent variable (output)
- domain of a function
- equation
- function
- graph
- independent variable (input)
- iterated function
- range of a function
- rate of change
- rule
- slope
- variable

Concepts to Know

- what a variable is
- what a function is and what it means for two quantities to have a functional relationship
- what the relationship between the independent and dependent variable is and what the domain and range of a function is
- the relationship between the function and its domain and range
- what rate of change is and what it means in functional relationships
- what it means to say that a function iterates
- what it means to find an optimal solution
- the relationship between functions and equations

Procedures to Know

- identifying functional relationships from tables, graphs, and symbols
- generating output values when given input values
- identifying a rule for a function from a table or a graph
- determining the domain and the range for a function
- using properties to solve equations for a variable
- finding rate of change of a function from a table, a graph, or an equation

Exercises & More Problems

Note: The exercises and problems at the end of the other chapters in this book are grouped into four categories: *Exercises, Critical Thinking, Extending the Activity,* and *Writing/Discussing.* The exercises and problems in this chapter have not been grouped in this way because so many of the exercises and problems belonging to different categories are associated with a table or a graph (e.g., items 1–5 are associated with the chart shown below). There are, however, *Writing/Discussing* tasks at the end of these exercises and problems.

In the chart below are estimates for the world population during the twentieth century:

Year	Population (in billions)
1900	1.7
1910	1.8
1920	1.9
1930	2.1
1940	2.3
1950	2.5
1960	3.0
1970	3.7
1980	4.5
1990	5.3

1. What was the rate of change of the world population, in people per year, between 1900 and 1910? Between 1900 and 1990?

2. What was the rate of change of the world population, in people per *minute*, between 1980 and 1990.

3. Based on the data, what do you predict will be the world population in 2000? Explain how you arrived at your estimate.

4. It has been predicted that the world population will "stabilize" around the year 2200, with a population of more than 11 billion. What factors do you think would contribute to this stabilization?

5. *(Optional)* For an interesting look at world population statistics, visit the World Wide Web at http://www.census.gov/cgi-bin/ipc/popclockw

The graph below shows the United States population from 1900 to 1990, as determined by the U.S. Census:

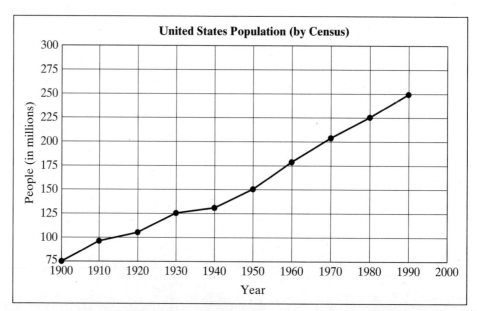

6. What was the rate of change of population, in people per year, from 1900 to 1990? Is this faster or slower than the rate of change of population from 1980 to 1990?

7. How can you tell, without doing any calculations, which of two time periods has the higher rate of change in population?

8. On average, according to the U.S. Census Bureau, there is a baby born every 8 seconds. At this rate, how many babies per year are born?

9. What do you predict will be the United States population in 2000?

10. Besides the birth and death rates, what other factors might contribute to the population of the United States?

11. Where on the graph do you see the lowest rate of change in population? What factors might have contributed to this decrease?

12. It feels colder outside when the wind is blowing than when it is not. The "windchill index" is a measure of the heat loss from the body when temperature and wind speed are combined. A formula for calculating windchill is as follows:

    $WC = 91.4 - (0.474677 - 0.020425 V + 0.303107 \sqrt{V})(91.4 - T)$,

    where WC is the windchill index in degrees Fahrenheit, V is the wind speed in miles per hour, and T is the air temperature in degrees Fahrenheit.

    Construct a Windchill Table with Air Temperature and Wind Speed as the variables. (A calculator may be a big help in solving this problem.)

13. Plot a line graph that shows the relationship between the length of one side of a square and the area of that square. Plot a line graph that shows the relationship between the radius of a circle and the area of that circle. How are the square graph and the circle graph similar? How are they different?

14. Plot a line graph that shows the relationship between the diameter of a circle (plotted on the horizontal axis) and the circumference of that circle (along the vertical axis). What is the slope of the resulting line?

15. The mass of an average American young person, in kilograms, is a function of the person's age in years. Below are some sample data. Choose reasonable scales for axes on a graph, and plot the given data. Then, write a sentence describing the pattern in the graph and what it says about the relation between the two variables.

Age (years)	0	2	4	6	8	10	12	14
Mass (kg)	3	11	15	20	26	31	38	49

    Source: *The World Almanac*, 1992.

16. You wish to construct a rectangular lot for your dog and to enclose it with a fence. Suppose that you want the lot to have an area of 400 square feet and that you have 240 feet of fence available but would like to have some fence left over.

    a. Suppose one side of the lot measured 10 feet. In all, how many feet of fence would you need to make the lot cover 400 square feet?

b. Make a table and a graph that show the relationship between the length of one side of the lot and the amount of fence you need for that lot.
c. What is the least amount of fencing needed to create the desired lot? Explain how you arrived at your solution.
d. Is it possible to create a lot with an area of 400 square feet that uses all of the available fence?

17. Suppose you have two natural numbers {1, 2, 3, ... } whose sum is 50. What is the largest possible *product* of these numbers? What is the smallest possible product?

18. Suppose you have *three* natural numbers whose sum is 20. What is the largest possible *product* of these numbers? What is the smallest possible product?

19. Suppose you have *some* natural numbers whose sum is 100. What is the largest possible product of these numbers?

20. At a grocery store, peanuts cost $2.29 per pound. Plot a graph of the relationship between weight (in pounds, along the horizontal axis) and cost (in dollars, along the vertical axis). The resulting graph will be a line. What will be the slope of that line? Explain how to find that slope without using a graph.

21. The Department of Mathematics is trying to decide between two new copying machines. One sells for $20,000 and costs $0.02 per copy to operate. The other sells for $17,500, but its operating costs are $0.025 per copy. The department decides to buy the more expensive machine. How many copies must the department faculty and staff make before the higher price is justified? Explain how you arrived at your solution.

22. A swimming pool with dimensions of 20 feet by 30 feet is surrounded by a sidewalk of uniform width x. Find the possible widths of the sidewalk if the total area of the sidewalk is to be greater than 200 square feet but less than 360 square feet.

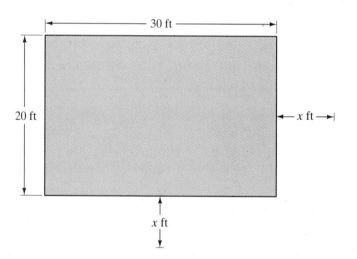

23. For # 23, write an equation that gives the area of the sidewalk in terms of x.

24. For # 23, write an inequality that is an algebraic representation of the problem situation.

25. A group is taking an outing, and they can rent a bus for $180 per day. The organizer of the trip decides to charge every member of the party an equal amount for the ride. How will the size of each person's contribution depend on the size of the group? Create a function to express the relationship.

26. A photographic service develops film for $2 (a fixed price for processing) plus 25¢ for each print. How does the cost of developing a film vary with the number of prints you want developed? Write down a function to express the relationship.

For each of the rules given below, complete the table of values.

27. Rule: Add five to the input and multiply the result by 3 to get the output.

Input	Output
1	
2	
3	
	42
x	
	a

28. Rule: Take 6% of the input, and add the result to the input to get the output.

Input	Output
50	
60	
70	
	424
x	
	a

29. Rule: Find the average of the input and 20 to get the output.

Input	Output
14	
−20	
20	
	125
x	
	a

30. Rule: Subtract 3 from the input, and divide the result into 20 to get the output.

Input	Output
2	
4	
6	
	1
x	
	a

In each of the following situations (#32–#34), you are given data relating two variables. In each case, choose reasonable scales for axes on a graph, and plot the given data. Then, write a sentence describing the pattern in the graph and what it says about the relation between the two variables.

31. The average weight of an American baby, in pounds, is a function of the baby's age in months. Here are some sample data.

Age (months)	Mass (lb)
0	7
3	13
6	17
9	20
12	22
15	24
18	25
21	26
24	27

32. The price of a bag of mangoes is a function of the number of mangoes bought. Here are some sample data.

Number of Mangoes	Price (dollars)
2	5
4	10
6	15
8	20
10	25
12	30
14	35
16	40

33. The displacement of a mass hung from an oscillating spring is a function of time. Here are some sample data.

time (sec)	displacement (in.)
0	1.0
1	0.4
2	−0.3
3	−0.5
4	−0.25
5	0.1
6	0.2
7	0.15
8	0
9	−0.1
10	−0.05
11	0
12	0.05

34. Think about functional relationships that you experience in everyday life. Many times we think of functional relationships where time is the independent variable. List two functional relationships involving time, where time is the *dependent* variable. Then sketch a graph of the function. For example, the amount of time it will take me to shovel my driveway can be thought of as a function of the depth of the snow.

Sketch a graph illustrating the relation between the variables in each of the following situations (#35–#37). Be sure to label your axes.

35. During the spring and summer, the height of a cornstalk is a function of growing time.

36. At a fruit stand, the price of a bag of apples is a function of the weight of the package.

37. When boiling water is poured over a tea bag, its temperature is a function of time.

38. Animal populations tend to rise and fall in cycles. Suppose the following data shows how squirrel populations in a central Pennsylvania city varied from 1975 to 1984. Plot a graph of this data. Then, write a headline and an opening paragraph for an article designed to inform readers about squirrel population patterns.

Year	'75	'76	'77	'78	'79	'80	'81	'82	'83	'84
Population	750	500	520	680	730	650	550	625	780	700

39. Find at least three different ways to fill in the next number in this series: 1, 2, 4, 7, ____. Explain how you arrived at each.

The Celsius scale for temperature was designed to make it easy to look at some common temperatures—those between the freezing point of water (known as 0° Celsius) and the boiling point of water (known as 100° Celsius). The Fahrenheit scale for temperature is set up differently, with the freezing point of water being 32° Fahrenheit and the boiling point of water known as 212° Fahrenheit. Knowing this, and knowing that that the relationship between degrees Celsius and degrees Fahrenheit is *linear,* answer the following questions (#40–#46).

40. If an object has a temperature of 50° Celsius, what is its temperature in degrees Fahrenheit? Explain how you found your answer.

41. In degrees Celsius, what range of temperatures might be appropriate for wearing shorts and a T-shirt while outside?

42. In degrees Celsius, what would a normal body temperature be?

43. At what temperature would the number of degrees Celsius be equivalent to the number of degrees Fahrenheit?

44. Find and describe a relationship between degrees Celsius and degrees Fahrenheit that could be used to help someone convert from one scale to the other.

45. Suppose someone suggests that an easy way to convert *approximately* from degrees Celsius to degrees Fahrenheit is to "double it and add 30," that is, multiply the number of degrees Celsius by 2 and add 30 to the product. How appropriate is this rule?

46. *Doubling a temperature in Celsius is the same as doubling a temperature in Fahrenheit.* Is this statement always true, sometimes true, or never true? Explain your reasoning.

A newspaper delivery person's weekly pay is a function of the number of papers delivered each day. Here are some sample data:

Papers delivered	50	100	150	200	300	500
Pay (dollars)	70	120	170	220	320	520

47. Plot a graph of this data. Suppose you were writing an advertisement to convince people to become newspaper carriers. How could you accurately describe, in a way that would be "attractive" to newspaper carriers, the relationship between the number of papers delivered and the pay the carrier receives?

48. According to wildlife experts, the rate at which crickets chirp is a function of the temperature; that is, $C = T - 40$, where C is the number of chirps every 15 sec and T is the temperature in degrees Fahrenheit.

   a. Sketch the graph of this function.
   b. Describe this function using as many descriptors as possible.
   c. Describe and explain what you think are the minimum and maximum values for this function.

49. Rebekah borrowed $250 from her mother to purchase a deluxe grass-catching lawn mower for her new lawn-mowing business. She created the table below to chart her summer business prospects. She wants her summer earnings to be no less than $1,000.

# of lawns mowed	Profit in $
0	−250
5	−200
10	−150
15	−100
20	−50
25	0
30	50

  a. How much is Rebekah charging per lawn?
  b. Develop a rule relating the number of lawns mowed and Rebekah's profit.
  c. How many lawns must Rebekah mow to reach her goal of $1,000?
  d. Suppose Rebekah decides to buy a less-expensive lawn mower for $195 instead of the $250 one. Develop a rule relating the number of lawns mowed and Rebekah's profit.
  e. How many lawns must Rebekah mow to reach her goal of $1,000, given that she starts with a $195 debt?

50. Notice that:

    $2^2 + 3^2 + 6^2 = 7^2$
    $3^2 + 4^2 + 12^2 = 13^2$, and
    $4^2 + 5^2 + 20^2 = 21^2$

  a. Is this a part of a general pattern? If so, what is the pattern?
  b. Write an equation for the general pattern.

51. According to the U.S. Office of Solid Waste of the Environmental Protection Agency, the annual production of municipal solid waste in the United States has grown significantly since 1960. In 1960, the United States produced 88 million tons of solid waste, and in 1984, about 176 million tons were produced. Assuming that the rate of growth in production of solid waste was constant from year to year, what was the annual rate of growth in solid waste during this period?

52. Africa is the world's least-urbanized, but most rapidly urbanizing, continent. A number of African cities have population growth rates of higher than 7%. Study the graph below, and then answer the questions that follow it.

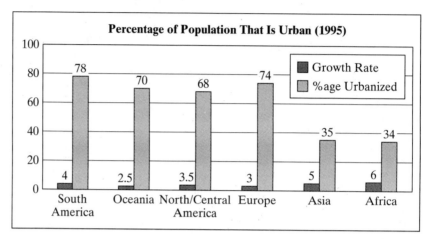

a. If the urban growth rates for Africa and Europe remain the same indefinitely, will Africa ever become more urbanized than Europe? Explain your answer.
b. Assuming that Oceania's growth rate has remained constant for the past 40 years, approximately what percentage of Oceania's population lived in urban areas in 1965?

Writing/Discussing

53. Make a concept map for *functions*, and explain the links and connections you made.

54. Describe the difference between a dependent (output) and independent (input) variable.

55. Write a paragraph explaining, in your own words, what you think a *function* is.

56. Is it possible for a set of data to have no rule that describes the relation between the input and output values? Make up a data set for which there is no rule. Is it always possible to find a rule to describe the relation between the input and output values if you are only given two input-output values? Explain.

57. Describe an everyday relationship that would *not* be a function, and explain why not.

58. Write a paragraph explaining, in your own words, what you think a *linear function* is.

59. Make a second concept map about functions. Write a reflection comparing your first and second maps and explaining your present understanding of functions.

61. Jamie and his sister Maria decided to go to a movie together. Jamie is afraid they will be late, so he starts out running. After a while, he grows tired and walks the rest of the way to the theater. Maria starts out walking but begins to run when she notices she might be late. They arrive at the theater at the same time. The graphs show the distances they have traveled from their home (vertical axis) and the time they have traveled (horizontal axis).

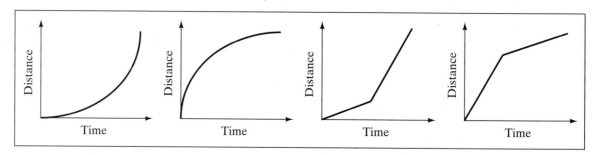

   a. Which graph best represents Maria's trip? Why do you think this?
   b. Which graph best represents Jamie's trip? Why do you think this?

Refer to the situation described in #61 above to answer #62–64.

62. Draw a graph of what Jamie's trip might look like if he ran half the distance and walked half the distance. Justify your answer.

63. Draw a graph of what Maria's trip might look like if she ran half the time and walked half the time. Justify your answer.

64. Who would arrive first at the movie theater if Jamie walked half the distance and ran the other half and Maria walked half the time and ran the other half? Justify your answer.

CHAPTER NINE

Geometry

CHAPTER OVERVIEW

Geometry is among the richest and oldest branches of mathematics. We think of geometry as the study of space experiences. This study focuses mainly on shapes as abstractions from the environment, which can be informally investigated and analyzed. In this chapter, you will explore two-dimensional shapes (although most of the ideas are equally valid for three-dimensional shapes as well). Particular attention is given to making conjectures and attempting to verify them—that is, to develop proofs for your conjectures.

BIG MATHEMATICAL IDEAS

Problem-solving strategies, shape and space, congruence, similarity, verifying, conjecturing, generalizing, decomposing

NCTM PRINCIPLES & STANDARDS LINKS

Geometry and Spatial Sense; Problem Solving; Reasoning; Communication; Connections; Representation

Activity **9.1** Communicating with Precise Language
9.2 Definitions: What is Necessary, and What is Sufficient?
9.3 Tangram Puzzles: Exploring Geometric Shapes
9.4 Constructing Geometric Relationships
9.5 Exploring Lines and Angles
9.6 Defining Angles and Lines
9.7 Exploring Polygons
9.8 Exploring Quadrilaterals
9.9 Properties of Quadrilaterals
9.10 Defining Triangles
9.11 Exploring Side Lengths in Triangles
9.12 Exploring Triangle Congruence
9.13 Exploring Triangle Similarity
9.14 More on Similarity
9.15 Constructing Proofs
9.16 Sums of Measures of Angles of Polygons
9.17 Spherical Geometry

ACTIVITY 9.1 Communicating with Precise Language

Topics in the Student Resource Handbook

9.1 **Two-dimensional Geometry Basics**

For this activity, you will need to work with a partner. You and your partner will sit across from each other and erect a barrier (e.g., an open book or a binder placed in a vertical position) so that neither of you can see what the other person is doing. Each partner should have a tangram set. One person forms a polygon with the tangrams and keeps it hidden from the partner. Then, this person describes the figure to the partner, and the partner tries to form the same figure, using only the verbal descriptions. The person who gives directions *cannot* describe the final figure (e.g., "It's a house"). The person following the directions may ask to have directions repeated or clarified, but neither person can look across or around the barrier unless the directions have been completed. No hand waving or signaling is allowed either! Once the person giving directions has completed them and the person following the directions is satisfied that he or she has followed them correctly, you may remove the barrier and compare the two figures. Switch roles and repeat.

ACTIVITY 9.2 Definitions: What is Necessary, and What is Sufficient?

FYI Topics in the Student Resource Handbook

9.1 **Two-dimensional Geometry Basics**
9.2 **Planes**
9.3 **Line Relationships**
9.4 **Line Segments**
9.5 **Rays and Angles**

What makes for a good definition? Writing a good definition involves carefully considering *necessary* and *sufficient* conditions and verifying that the definition says what you think it says. In this activity, your instructor will lead you in discussing definitions for some geometric terms. Use this page to record the class's progress toward good definitions of the geometric concepts you discuss.

ACTIVITY 9.3 Tangram Puzzles: Exploring Geometric Shapes

FYI Topics in the Student Resource Handbook
9.7 Polygons

The seven polygons below comprise the ancient Chinese tangram puzzle.

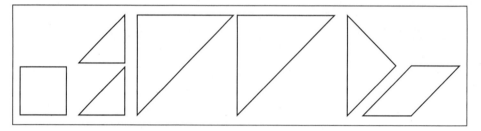

The puzzle involves manipulating the square, the parallelogram, and five triangles into various silhouette patterns of people, animals, and other figures. For example, the silhouette of the person below can be formed with the seven tangram pieces.

1. Use your tangram pieces to form this figure.

Is it possible to form a square using all seven tangram pieces? Is it possible to form a trapezoid using only five of the pieces? The answers to such questions may not be immediately obvious. By exploring the sizes and shapes of the pieces, you may make some discoveries that can help you to complete the chart below. As you explore, if you discover any relationships among the pieces, write them down so that you may share them with others.

2. In the chart below, use your tangram pieces to determine whether the given shapes can be formed. Record your solutions so that you may later share them.

	Using only...						
Can you form a...	1 piece	2 pieces	3 pieces	4 pieces	5 pieces	6 pieces	7 pieces
Square							
Nonsquare rectangle							
Trapezoid							

3. A student made the claim that if two figures are made using exactly the same tangram pieces, then they must have the same area. Is this claim true or false? How can you convince others?

4. Another student claimed that if two figures are made of the exact same tangram pieces, then they must have the same perimeter. Is this claim true or false? How can you convince others?

ACTIVITY 9.4 Constructing Geometric Relationships

FYI Topics in the Student Resource Handbook
- 9.14 Lines and Angles Constructions
- 9.15 Bisector Constructions
- 9.16 Perpendicular-Line-Through-a-Point Constructions
- 9.17 Parallel-Line Construction

For generations, geometers were interested in knowing which geometric figures could be constructed given some geometric elements such as points, segments, lines, angles, or circles and using *only* a straightedge (no markings on it) and a collapsible compass.

The *Student Resource Handbook* has instructions for performing some basic compass and straightedge constructions. It will be helpful to read them before making the following constructions.

1. Construct a segment congruent to $\overline{PQ}$ on the given line *m*.

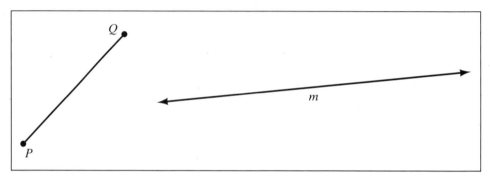

2. Construct a segment congruent to $2\overline{PQ}$ on the given line *n*, where $\overline{PQ}$ is as in question 1.

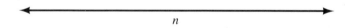

3. Construct an angle congruent to $2 \cdot m\angle A$ below which has $\overrightarrow{PX}$ as one of its sides.

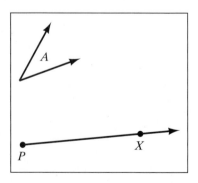

4. Construct a line through point *P* that is parallel to the given line *k*.

5. Using only a compass and a straightedge, construct a right triangle with one of its legs twice the length of the other.

6. Using only a compass and straightedge, construct an equilateral triangle *ABC* (make each side approximately 4 inches—estimate this length) with base *BC*. Now, bisect the two base angles of the triangle (<*B* and <*C*), and extend these angle bisectors until they meet at point *D*. Connect points *B* and *D* and points *C* and *D*. Make a conjecture about the shape of triangle *BCD*. Check your conjecture by measuring.

ACTIVITY 9.5 Exploring Lines and Angles

FYI Topics in the Student Resource Handbook

- 9.1 Two-dimensional Geometry Basics
- 9.2 Planes
- 9.3 Line Relationships
- 9.4 Line Segments
- 9.5 Rays and Angles
- 9.6 Angle Relationships

For this activity, your instructor will give you a paper triangle. Color each angle of the triangle a different color (e.g., red, yellow, and green). Draw a straight baseline near the middle of a blank sheet of paper. Place one side of the triangle on the baseline at one end of it. Trace the other two sides of the triangle.

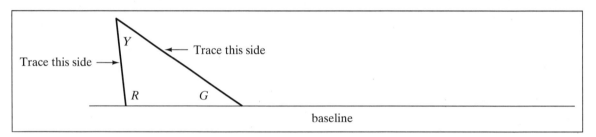

Now, rotate the triangle so that the angle marked Y fits snugly against the baseline and the side common to angles Y and G fits against its tracing. Then trace the side common to angles Y, and R.

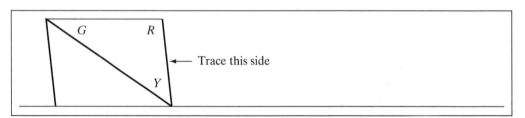

Continue to rotate the triangle, and trace the sides of the angle Y to create a figure with "saw teeth." Color each angle Y as you make it.

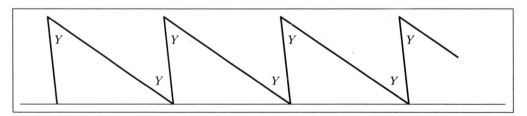

1. Alternating sides of the saw "teeth" form _____ lines, and the angles within the saws are _____.

Activity 9.5 *Exploring Lines and Angles* **363**

To create a figure with "ladder rungs," draw another baseline, this time toward the lower end of your blank paper. Place one side of the triangle along the baseline at one end of it. Of the two sides not on the baseline, trace the side closest to the left end of it. Mark the point where the thiird vertex meets the baseline.

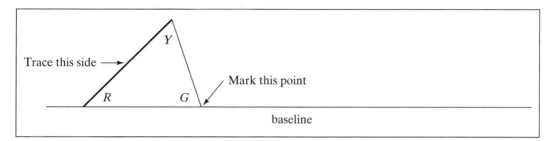

Now, slide the triangle along the baseline until the vertex of angle *R* lies on the marked point and repeat—trace the same side, and mark a point using the same vertex as before.

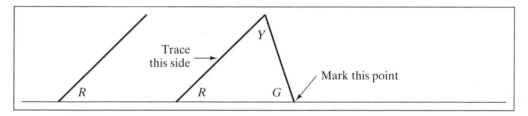

Keep sliding the triangle and tracing the side of angle *R* that is not on the baseline. These sides appear as the "rungs" of a ladder. Color each angle *R* as it is made.

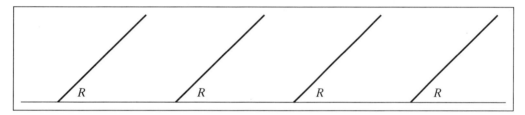

2. The rungs of the ladder form _____ lines, and the angles on the same side of each rung are _____ .

Interior Angles

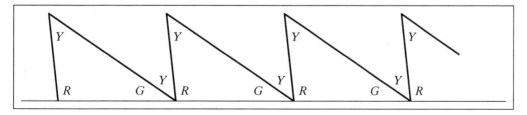

3. Using both the saws and the ladder patterns, one gets the above figure. Make a conjecture about the interior angles of a triangle.

Angles Formed by the Intersection of Two Lines

Using a straightedge, draw two intersecting lines. Label each of the four angles, and use those labels to answer the following questions.

4. Using a protractor, measure each of the angles, and record your measurements below.

5. What do you notice about the measures of the angles opposite each other?

6. What do you notice about the measures of the angles adjacent to each other?

7. Do you think these relationships will always be true whenever two straight lines intersect each other? Why or why not?

Angles Formed by the Intersection of a Transversal Line and Two Other Lines

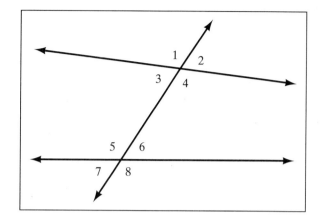

8. <1 and <5 are a pair of corresponding angles. Find three more pairs of corresponding angles.

9. <3 and <6 are a pair of alternate interior angles. Find another pair of alternate interior angles.

10. <2 and <7 are a pair of alternate exterior angles. Find another pair of alternate exterior angles.

11. Based on the saws-and-ladders activity, make conjectures about corresponding angles and alternate interior angles when a transversal line intersects two parallel lines.

12. Do you think these relationships will always be true whenever two parallel lines are cut by a transversal? Why or why not?

ACTIVITY 9.6 Defining Angles and Lines

FYI Topics in the Student Resource Handbook
- 9.3 Line Relationships
- 9.4 Line Segments
- 9.5 Rays and Angles
- 9.6 Angle Relationships

Use the examples and nonexamples below to define the following types of angles and lines.

1. Define *right angle*

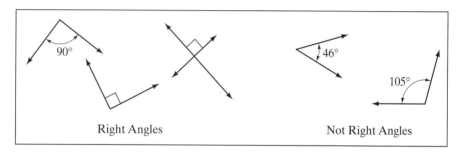

(Note: a small square in the corner of an angle indicates that its measure is 90°)

2. Define *acute angle*

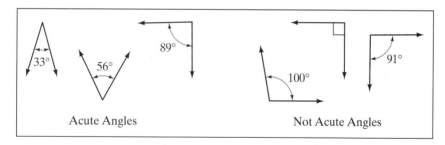

3. Define *obtuse angle*

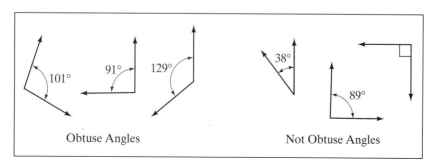

4. Define *pair of adjacent angles*

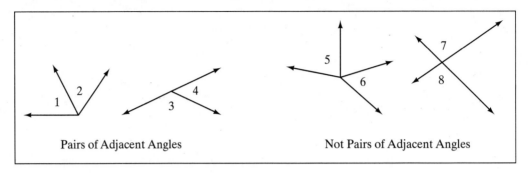

5. Define *pair of complementary angles*

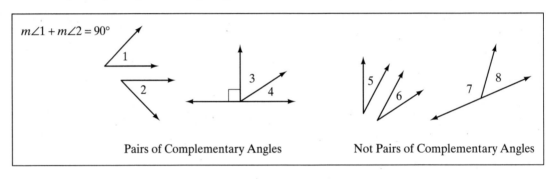

6. Define *pair of supplementary angles*

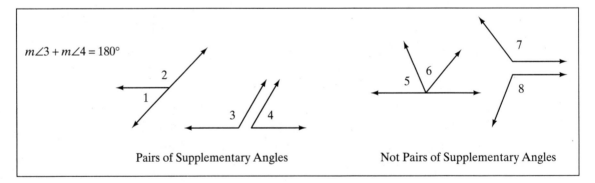

7. Define *pair of vertical angles*

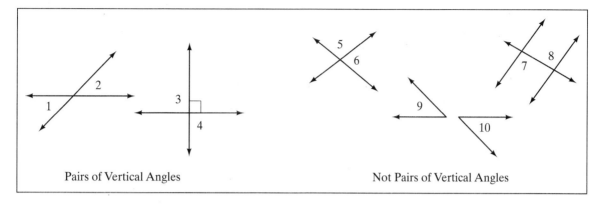

Pairs of Vertical Angles

Not Pairs of Vertical Angles

8. Define *linear pair of angles*

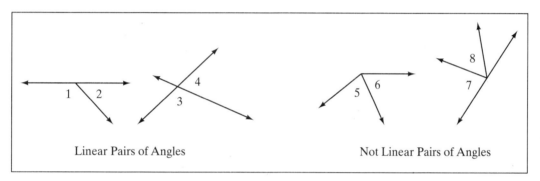

Linear Pairs of Angles

Not Linear Pairs of Angles

9. Define *collinear points*

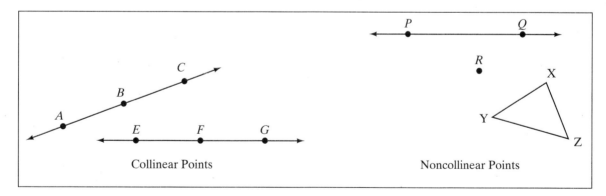

Collinear Points

Noncollinear Points

10. Define *coplanar points*

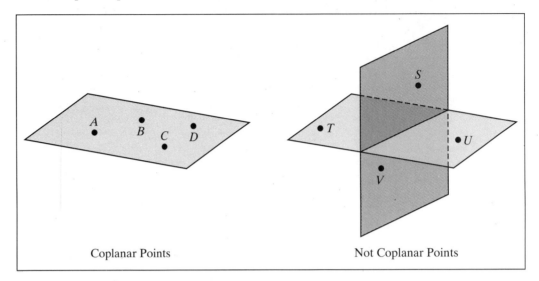

11. Define *parallel lines*

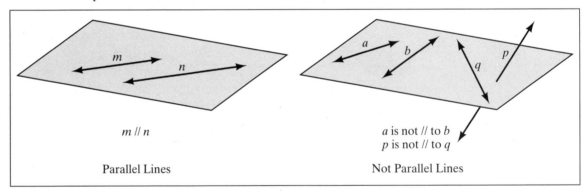

(note: // means "is parallel to")

12. Define *perpendicular lines*

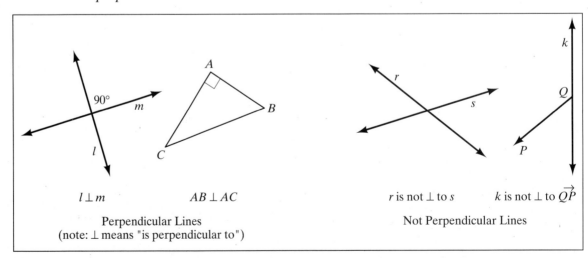

ACTIVITY 9.7 Exploring Polygons

FYI Topics in the Student Resource Handbook
9.7 **Polygons**
9.8 **Regular Polygons**

1. Given below are some examples and nonexamples of polygons.

   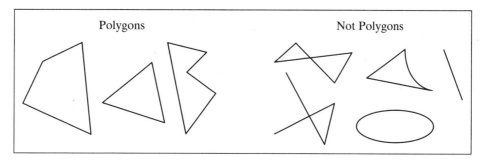

   a. Give a definition of a polygon based on these examples and nonexamples.

   b. Quadrilaterals and triangles are examples of polygons. What are some other polygons?

   c. A *regular* polygon is one with all sides congruent and all angles congruent. What do we call a regular quadrilateral?

d. The first two polygon examples shown at the top of page 371 are *convex* polygons. The third is a *concave* polygon.

   i. Draw a concave quadrilateral.

   ii. Which polygon is always convex?

ACTIVITY 9.8 Exploring Quadrilaterals

FYI Topics in the Student Resource Handbook
9.10 **Quadrilaterals**

A quadrilateral is a polygon with four sides. Look carefully at the convex quadrilaterals below, and answer the following questions.

1. Which quadrilaterals have at least one pair of parallel sides?

2. Which quadrilaterals have two pairs of parallel sides?

3. Which quadrilaterals have two distinct pairs of consecutive angles congruent?

4. Which quadrilaterals have two distinct pairs of consecutive sides congruent?

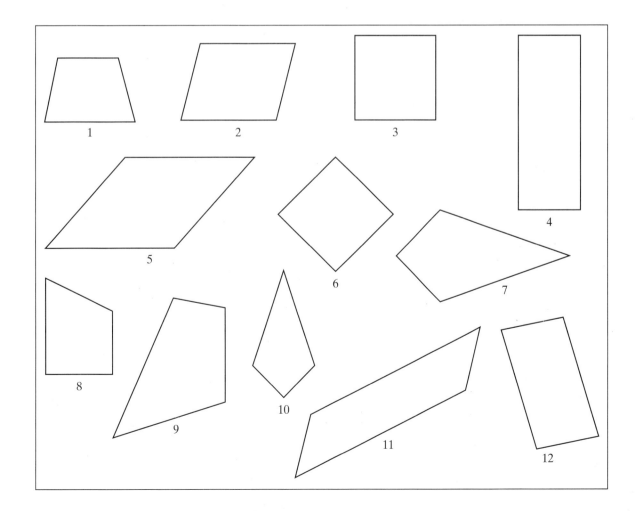

5. Match the 12 quadrilaterals with the names below. For each name, list all the quadrilateral numbers that fit this name. Note that we will define a *trapezoid* as a quadrilateral with at least one pair of parallel sides (instead of other definitions that require exactly one pair of parallel sides).

   a. parallelogram

   b. square

   c. polygon

   d. trapezoid

   e. rectangle

   f. kite

   g. rhombus

   h. isosceles trapezoid

6. Fill in the blanks, using "All" or "Some."

   a. _____ parallelograms are rhombi.

   b. _____ kites are squares.

   c. _____ rectangles are trapezoids.

   d. _____ squares are quadrilaterals.

   e. _____ isosceles trapezoids are parallelograms.

ACTIVITY 9.9 Properties of Quadrilaterals

FYI Topics in the Student Resource Handbook
9.10 **Quadrilaterals**

1. On the next page, complete the table that lists different properties that quadrilaterals could possess. Some of these you may already know. Others you may check using a tool such as the *Geometer's Sketchpad*™, paper-and-pencil constructions, or manipulatives that your instructor has available. Make sure you understand what the various terms mean. Discuss these in your group, and ask the instructor if you are still unsure.

2. After you have completed the table, use it to complete the tree diagram below, relating the following quadrilaterals: isosceles trapezoid, kite, parallelogram, rectangle, rhombus, square, trapezoid.

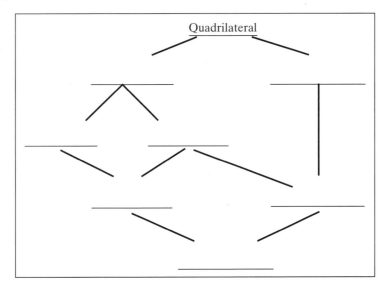

3. Use the table and your tree diagram to consider the number of side properties of each quadrilateral. What happens to the number of side properties as you move down the tree diagram?

4. Consider the number of diagonal properties of each quadrilateral. What happens to the number of diagonal properties as you move down the tree diagram?

5. Describe any connections you see between the relationships among the quadrilaterals and the properties they possess.

PROPERTIES	QUADRILATERALS						
	Parallelogram	Rectangle	Rhombus	Square	Trapezoid	Isosceles Trapezoid	Kite
Side Properties							
4 sides	X	X	X	X	X	X	X
At least 1 pair of parallel sides							
2 pairs of parallel sides							
All sides congruent							
At least 1 pair of opposite sides congruent							
Opposite sides congruent							
2 pairs of congruent adjacent sides							
Angle Properties							
Interior angle sum = 360°							
All angles are right angles							
Opposite angles are congruent							
Adjacent angles are supplementary							
Diagonal Properties							
Diagonals bisect each other							
Diagonals are congruent							
Diagonals are perpendicular							
Diagonals bisect vertex angles							
1 diagonal forms 2 congruent triangles							
Diagonals form 4 congruent triangles							
Symmetric Properties							
Number of lines of symmetry							
Rotational Symmetry							

ACTIVITY 9.10 Defining Triangles

FYI Topics in the Student Resource Handbook

9.9 Triangles

A triangle is a polygon with three sides. Based on the examples and nonexamples given below, define the following geometric terms.

1. Define *right triangle*.

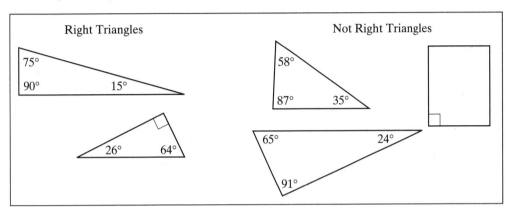

2. Define *acute triangle*.

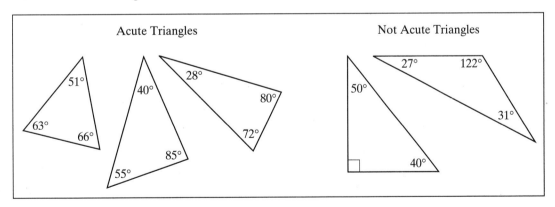

3. Define *obtuse triangle*.

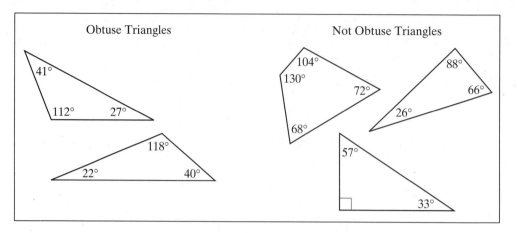

4. Define *scalene triangle*.

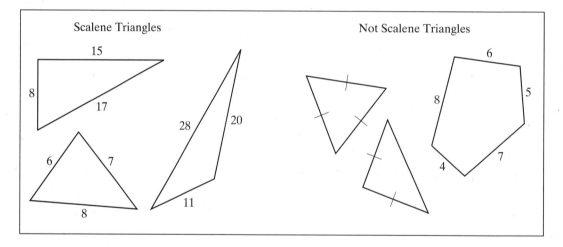

5. Define *isosceles triangle*.

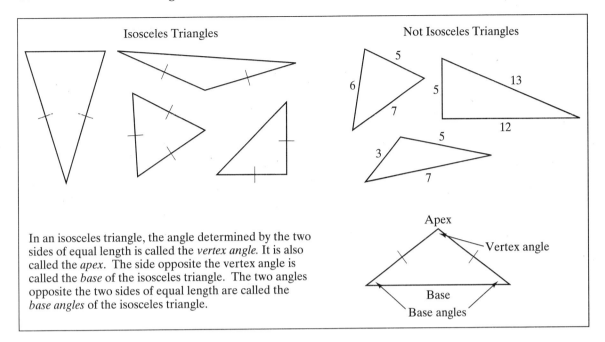

In an isosceles triangle, the angle determined by the two sides of equal length is called the *vertex angle*. It is also called the *apex*. The side opposite the vertex angle is called the *base* of the isosceles triangle. The two angles opposite the two sides of equal length are called the *base angles* of the isosceles triangle.

6. Define *equilateral triangle*.

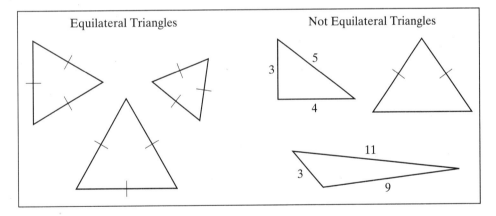

7. Define *median of a triangle*.

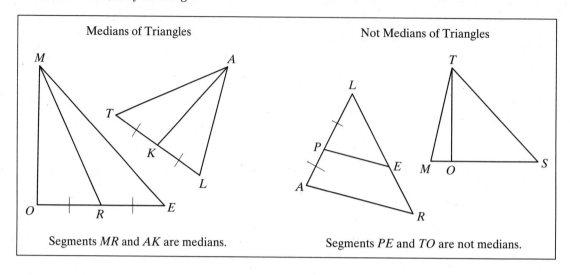

8. Define *altitude of a triangle*.

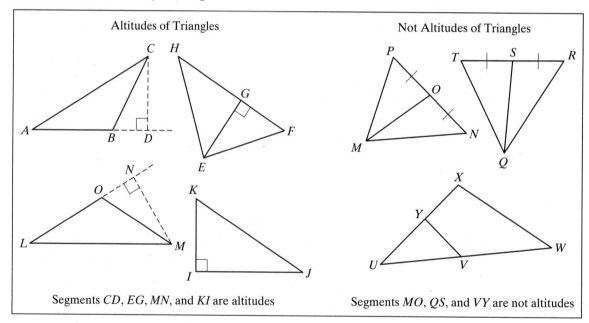

ACTIVITY 9.11 Exploring Side Lengths in Triangles

FYI Topics in the Student Resource Handbook
9.9 Triangles

Try the following constructions on blank paper, using only a compass and ruler.

1. Construct a triangle with side lengths 3 in., 5 in., and 2 in. What happened?

2. Construct a triangle with side lengths 6 in., 2 in., and 3 in. What happened?

3. Construct a triangle with side lengths 3 cm, 10 cm, and 4 cm. What happened?

4. Construct a triangle with side lengths 7 cm, 4 cm, and 6 cm. What happened?

5. Use the above to fill in the values in the chart below.

Side lengths a, b, c	$a + b$	$b + c$	$c + a$	What happened?

6. What do you notice when you compare the length of the third side to the sum of the lengths of the other two sides? State your observation in the form of a theorem in two different ways: in a general form as a statement about triangles and in an "if ... , then ... " form.

ACTIVITY 9.12 *Exploring Triangle Congruence*

Topics in the Student Resource Handbook
9.12 **Congruence**

1. Two things are said to be congruent if they are identical in all their parts. How would you define *congruent triangles*?

2. Using a ruler and compass, on a blank sheet of paper, construct a triangle whose sides measure 2 in., 3 in., and 4 in. On a second sheet of paper, construct in a different way another triangle with the same side lengths as the first. When you have finished, try to fit one triangle over the other (holding the two sheets of paper up to the light might help.) What do you find?

3. Use your triangle with sides of length 2 in., 3 in., and 4 in. Can you construct a triangle with sides of length 2 in., 3 in., and x in that is not congruent to the first? What did you take as x? What could you take as x?

4. In #2, what were the conditions on the two triangles with which you started? Generalize these conditions so that they can be used as the hypothesis of a theorem. (Hint: The hypothesis is the "if" part of the theorem.)

5. In #2, what happened when you compared the two triangles? Will the same thing happen under the generalized conditions you stated in #4? Call this the conclusion of your theorem. (Hint: The conclusion is the "then" part of the theorem.)

6. Using your answers to #4 and #5, state your theorem below, and give it a descriptive name.

7. Suppose we try a slightly different approach to making triangles. You are given the lengths of two sides of a triangle—2 in. and 4 in.—and the measure of the angle formed by the two sides (called the included angle)—30°. Using only your ruler and protractor, construct, in different ways on two sheets of blank paper, two triangles with these measurements. What do you notice when you compare the two triangles?

8. Make two more triangles where the sides are 3 in. and 7 in. and the included angle is 45°. What do you notice when you compare them?

9. This time, try something just a little different. Construct one triangle using lengths 3 in. and 5 in. and included angle 45° and the other triangle using lengths 3 in. and 5 in. and included angle 48°. Now compare the two triangles. What happened? Why?

10. State another theorem that gives necessary and sufficient conditions for the congruence of two triangles. Name your theorem.

ACTIVITY 9.13 Exploring Triangle Similarity

FYI Topics in the Student Resource Handbook
9.23 **Similarity**

1. In this activity, you will use the top $\triangle ABC$ and point P shown on page 387. Follow the directions below to construct a new triangle from $\triangle ABC$.

   - Construct the ray with endpoint P through point A.
   - Construct the ray with endpoint P through point B.
   - Construct the ray with endpoint P through point C.
   - Set your compass opening to the length PA.
   - Starting at P, mark off two lengths of PA along the ray PA.
   - Label this endpoint D.
   - Set your compass opening to the length PB.
   - Starting at P, mark off two lengths of PB along the ray PB.
   - Label this endpoint E.
   - Set your compass opening to the length PC.
   - Starting at P, mark off two lengths of PC along the ray PC.
   - Label this endpoint F.
   - Connect the points D, E, and F.

2. Make conjectures as to what the relationships are between $\triangle ABC$ and $\triangle DEF$.

3. Explain why you think the conjectures you made in #2 are true.

4. Use the second $\triangle ABC$ on page 387 after the first $\triangle ABC$. This time notice that point P is inside $\triangle ABC$. Repeat the steps in #1 above. What happens?

Activity 9.13 *Exploring Triangle Similarity* **387**

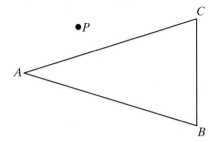

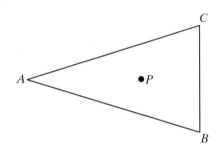

ACTIVITY 9.14 More On Similarity

FYI Topics in the Student Resource Handbook
9.23 **Similarity**

1. In Activity 9.13, you explored similar triangles. What does it mean for one triangle to be similar to another triangle? What does it mean for one object to be similar to another? What are some objects in your everyday life that are similar to each other?

2. Some photocopy machines can be programmed to enlarge or reduce the original, and both copies are similar to the original. An enlargement or reduction is called a *dilation*. Thus, a similar figure can be obtained by dilating the original. A machine that is common in classrooms dilates an original. What is it? Does it dilate by enlarging or reducing, or both?

3. Your instructor will now lead the class in an activity that will enable you to discover the existence of the center of dilation and scale of dilation and how they are related. Use the rest of this page to record the class activity, which will include a discussion of some properties of similar figures.

ACTIVITY 9.15 Constructing Proofs

FYI Topics in the Student Resource Handbook

9.1 Two-dimensional Geometry Basics
9.26 Three-dimensional Geometry Basics

Prove the following statements.

1. In a triangle, angles opposite congruent sides are congruent.

2. If both pairs of opposite sides of a quadrilateral are congruent, then the quadrilateral is a parallelogram.

3. In any triangle, the line segment formed by connecting the midpoints of two sides is parallel to the third side.

4. Draw a square *ABCD*. Draw the line segments connecting the midpoints of adjacent sides. Label the new quadrilateral *EFGH*. Prove that quadrilateral *EFGH* is also a square.

ACTIVITY 9.16 Sums of Measures of Angles of Polygons

Topics in the Student Resource Handbook

9.7 Polygons
9.8 Regular Polygons

1. How could you show that the sum of the measures of the angles of any triangle is always 180°? How could you prove it? Write the proof.

2. In any convex polygon with three or more sides, there is a relationship between the number of sides and the sum of the measures of the interior angles. Find the relationship. Justify your answer. Completing the table on the next page will help you find this relationship.

3. Is there any difference in the sum of the angle measurements in regular polygons and in nonregular polygons? Why do you think this is the case? Justify your answer.

4. Do these relationships hold for concave polygons? Why or why not?

Polygon	# of Sides	Sum of Angle Measurements	Measurement of Each Angle If Polygon Is Regular
Triangle		180°	
Square		360°	
Pentagon	5		
Hexagon	6		
Heptagon	7		
Octagon	8		
Nonagon	9		
Decagon	10		
11-gon			
Dodecagon	12		
13-gon			
57-gon			
89-gon			
n-gon			

ACTIVITY 9.17 Spherical Geometry

FYI Topics in the Student Resource Handbook
9.26 Three-dimensional Geometry Basics

True or False:

Through a point not on a line, there is exactly one line perpendicular to the line. _____

If two lines intersect, their intersection contains only one point. _____

The sum of the angle measures of a triangle is 180°. _____

These statements are all true in the geometry developed by Euclid. They are all false in *spherical geometry*. In spherical geometry, a flat surface is not used but rather the surface of a sphere. A line in spherical geometry is a great circle of the sphere. What is a great circle?
_____

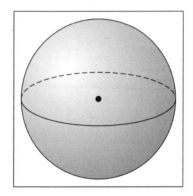

Do spherical lines have length? _____ What is the length of a line in spherical geometry?
_____

A point in spherical geometry is any point on the surface of a sphere.
How many points of intersection are there for two spherical lines?
_____ Why? _____

In the figure at the top of page 394, points A, B, and C are on the same great circle. Are A, B, and C collinear? _____

Which of the three points is between the other two? _____

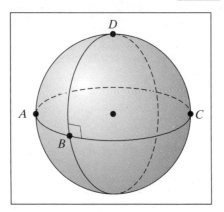

Suppose the spherical line through D is perpendicular to line AB. Are there other lines perpendicular to line AB? _____ If so, how many? _____ Illustrate this in the diagram.

Are there any lines parallel to line AB? _____

Now, consider special triangles. In spherical triangle ABC below, is the sum of the angle measures 180°? _____

Is it greater or less than 180°? _____ How many right angles can a triangle have in Euclidean geometry? _____

How many right angles can a triangle have in spherical geometry? _____ Why? _____

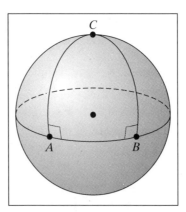

A triangle in Euclidean geometry can have only one obtuse angle. How many obtuse angles can a spherical triangle have? _____ Why? _____

An obtuse angle has a measure greater than 90° and less than 180°. Thus, the sum of the angle measures of a spherical triangle is less than _____.

Things to Know from Chapter 9

Words to Know

- adjacent angles
- alternate exterior angles
- angle
- collinear points
- concave polygon
- convex polygon
- corresponding parts
- exterior angles
- interior angles
- parallel lines
- pentagon
- polygon
- ray
- regular polygon
- similarity
- supplementary angles
- triangle
- vertical angles
- alternate interior angles
- altitude
- circle
- complementary angles
- congruence
- coplanar points
- definition (necessary and sufficient conditions)
- hexagon
- octagon
- parallelogram
- perpendicular lines
- quadrilateral
- rectangle
- rhombus
- square
- trapezoid
- vertex

Concepts to Know

- what it means to define a term with necessary and sufficient conditions
- what it means to construct a geometric figure
- the relationships among various polygons (e.g., quadrilaterals, triangles)
- what it means to prove something
- the relationship among the side lengths of a triangle
- what congruence is, what similarity is, and the relationships between congruent and similar figures
- the relationship among the sum of the measures of the angles in a polygon and its number of sides
- what spherical geometry is and what its relationship to Euclidean geometry is

Procedures to Know

- writing definitions of terms with necessary and sufficient conditions
- recognizing and using various geometric figures and shapes
- constructing basic geometric figures
- classifying polygons according to their properties
- determining whether three given segment lengths could be used to form a triangle
- determining when two figures are congruent
- determining when two figures are similar
- using properties of figures to find angle measures and/or side lengths
- determining the sum of the measures of the angles in a polygon
- determining the measure of an angle in a regular polygon

Exercises & More Problems

Exercises

1. Determine how many lines are determined by the following:
   a. four noncollinear points
   b. five points, where three are noncollinear

2. A classmate claims that if any two planes that do not intersect are parallel, then any two lines that do not intersect should also be parallel. What do you think?

3. Give a mathematical explanation of why a three-legged stool is always stable and a four-legged stool sometimes rocks.

4. Is it possible that a line is perpendicular to one line in a plane but is not perpendicular to the plane? Explain.

5. Determine whether the following triangles can exist (draw a picture if one does exist):
   a. an right scalene triangle
   b. an obtuse isosceles triangle
   c. an acute scalene triangle
   d. a scalene isosceles triangle

6. Determine which of the following are either congruent or supplementary when a transversal intersects two parallel lines:
   a. alternate exterior angles
   b. same side interior angles
   c. corresponding angles
   d. vertical angles
   e. adjacent angles
   f. alternate interior angles

7. If two angles are complementary and the ratio of their measures is 5:4, then what are their angle measures?

8. Using any of the three constructions shown in Activity 9.4, construct an isosceles triangle that has the following properties:

   a. the length of its base is congruent to $\overline{AB}$ and its base angles are congruent to $\angle Q$

   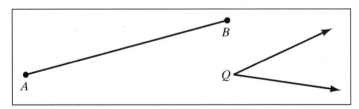

   b. the length of its base is congruent to AB above and its equal sides are of length $2 \cdot AB$

9. Using only a compass and a straightedge, construct the following quadrilaterals:
   a. a square
   b. a rectangle that is not square
   c. a parallelogram that is not a rectangle
   d. an isosceles trapezoid that is not a rectangle
   e. a rhombus
   f. a kite that is not a rhombus

10. a. Complete the following to obtain true statements. Remember that a theorem has no exceptions.

    i. A quadrilateral with four congruent sides is a _____.

    ii. A quadrilateral with three congruent sides is a _____.

    b. Consider statements i and ii below. If true, explain. If false, find a counterexample.

    i. A quadrilateral with all angles congruent is a square.
    ii. A quadrilateral with all angles congruent is a rectangle.

11. Sketch and carefully label each figure:

    a. an acute isosceles triangle
    b. an obtuse isosceles triangle
    c. a right isosceles triangle
    d. a scalene triangle with median
    e. an equilateral triangle with altitude

12. Draw a triangle, and construct one of its altitudes.

13. What is the minimum information needed to determine congruency for each of the following? Explain.

    a. two squares　　　b. two rectangles　　　c. two parallelograms

14. Which of the following are always similar? Explain.

    a. any two rectangles　　　b. any two circles　　　c. any two regular polygons

15. True or False?

    a. Any two isosceles triangles are similar.
    b. Any two equilateral triangles are similar.
    c. Any two isosceles right triangles are similar.
    d. A triangle has exactly one median.
    e. A triangle has at least three altitudes.

16. True or False? If false, give a counterexample.

    a. All circles are similar.
    b. All circles are congruent.
    c. All rectangles are similar.
    d. All squares are similar.
    e. All similar polygons are congruent.
    f. All congruent polygons are similar.

17. Determine whether a 5×8 rectangle is similar to any of the following:

    a. 10×16　　　b. 15×32　　　c. 30×54　　　d. 20×32

18. Shown below is a series of six rectangles that have been nested according to size, with the longer edge at the bottom and the lower left corners aligned.

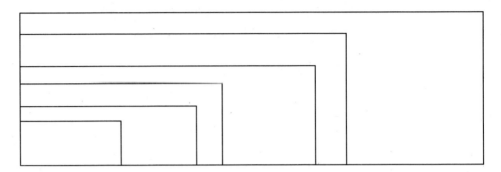

a. Which rectangles do you think are similar? Carefully measure ratios of corresponding sides, and check your guess.
b. How do those you identified as similar nest together? What might be a test for determining similarity of rectangles?

19. Write a definition for the terms below. Remember to use necessary and sufficient conditions.

a. A rectangle is a quadrilateral with …
b. A rhombus is a quadrilateral with …
c. A square is a quadrilateral with …

20. If quadrilateral $ABCD$ is a parallelogram, and $BF = 6$, $FC = 3$ and $BD = 7.5$, find BE and ED. Explain your answer.

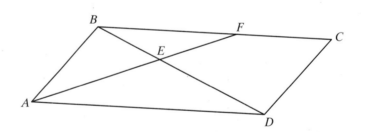

21. Using the chart below, write a formula for $a°$ in terms of s.

s	$a°$
3	150
4	300
5	450
6	600
.	
.	
.	
n	

22. Find the sum of the interior angles of the polygon illustrated below. Justify your answer.

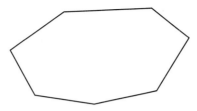

23. Consider the sets of quadrilaterals (Q): parallelograms (P), trapezoids (T), rectangles (R), rhombi (D), and squares (S). Draw a Venn diagram to illustrate the relationships among these sets of quadrilaterals, and explain your diagram.

24. Consider the following diagram:

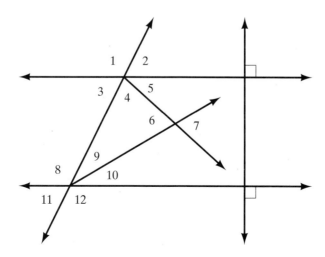

   a. What is the relationship between <1 and <3?

b. What is the relationship between <6 and <7?
c. What is the relationship between <2 and <11?
d. What is the relationship between <8 and the sum of <4 and <5?

25. Given the figure below, with segment *AX* parallel to segment *DY*, find the following:

    a. m<1
    b. m<3
    c. m<4
    d. m<2
    e. m<5

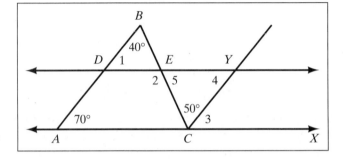

26. Find the measure of each interior angle of a regular hexagon. Explain your procedure.

27. Find the measure of each interior angle of a regular octagon. Explain your procedure.

Critical Thinking

28. Into how many regions would a plane be separated by two parallel lines? By two intersecting lines?

29. Is it possible for a line to be parallel to another line in a plane but not be parallel to the plane itself? Explain.

30. List the similarities between rays, segments, and lines.

31. List the similarities between the terms *collinear* and *coplanar*.

32. Can a line be perpendicular to two distinct lines in a plane and not be perpendicular to the plane? Explain.

33. If a line not in a given plane is perpendicular to two distinct nonparallel lines in the plane, is the line necessarily perpendicular to the plane? Explain.

34. Justify why this is a good or bad definition of *adjacent angles*: two angles with a common side.

35. Justify why this is a good or bad definition of a *square*: a parallelogram with one right angle.

36. In the following diagram, a student claims that polygon *ABCD* is a parallelogram if m<1 = m<2. Is she correct? Justify your answer.

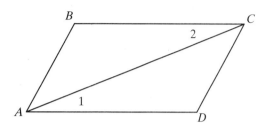

37. Is it possible for two triangles to be congruent but not similar? Explain.

38. Which polygons are always similar regardless of the dimensions chosen?

39. If two sides of one triangle were congruent to two sides of another triangle, then what other conditions must also be satisfied for the two triangles to be congruent?

40. Suppose that in the figure below, segment *AB* is parallel to segment *DE*, *BF* is parallel to segment *CD*, and m<*ABF* = 90°, *AB* = 6, *BF* = 8, and *DE* = 3.

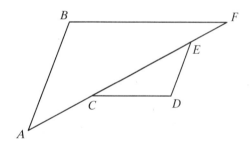

   a. What is m<*CDE*? Explain.
   b. What is the length of segment *CD*? Explain.
   c. What are the lengths of segments *EC* and *AF*? Explain.

41. A triangle *ABC* has the following side lengths: side opposite <*A* is length *a*, side opposite <*B* is length *b* and side opposite <*C* is length *c*.

   a. If $a > b > c$, what is the relationship between m<*A*, m<*B* and m<*C*?
   b. If $a = b$ and each is less than *c*, what is the relationship between m<*A*, m<*B* and m<*C*?
   c. If m<*B* = m<*C* and each is greater than m<*A*, what is the relationship between *a*, *b*, and *c*?

42. Determine whether each of the following statements are SOMETIMES, ALWAYS, or NEVER TRUE:

   a. If two triangles are congruent, they are similar.
   b. If two triangles are similar, they are congruent.
   c. If two triangles are both equilateral, they are similar.
   d. If two triangles are both isosceles, they are similar.

43. Reflect on the following statement and then answer parts a and b: If a trapezoid is cut into four triangles by its two diagonals, at least two of the four triangles are congruent.
    a. Is the statement ALWAYS, SOMETIMES, or NEVER TRUE.
    b. If your answer to a. is ALWAYS or NEVER TRUE, give a reason why. If your answer is SOMETIMES, give an example of when it is true and an example of when it is false.

44. Is it sufficient to say that two triangles are similar if they have two pairs of congruent angles? Explain.

45. Is it sufficient to say that two triangles are similar if they have all pairs of corresponding sides in the same ratio? Explain.

46. Is it sufficient to say that two rectangles are similar if they have two pairs of congruent angles? Explain.

47. Is it sufficient to say that two rectangles are similar if they have four pairs of congruent angles? Explain.

48. Is it sufficient to say that two rectangles are similar if they have all pairs of corresponding sides in the same ratio? Explain.

49. a. Show by computing ratios that two triangles with sides length 4 in., 3 in., 2 in., and 6 in., 4.5 in., and 3 in. respectively have proportionate sides.
    b. Using only a compass and a ruler, construct the above triangles. Using a protractor, measure the interior angles.
    c. Does your conjecture in #46 hold true?

50. a. Using only a straightedge and a protractor, construct two triangles that have the same interior angles—45°, 35°, and 100° as follows: Start with base sides of unequal lengths, and using your protractor and a straightedge, construct different pairs of the above three angles at the two vertices formed by the ends of the base. Extend the arms of the angles until they meet at the third vertex. (Measure this angle—it should be equal to the missing third angle. Why?)
    b. Using your ruler, carefully measure the sides of each triangle, and calculate the ratios of sides opposite congruent angles in the two triangles.
    c. Does your conjecture in #45 hold true?

51. In each of the following, find x and y if possible.

    a.

    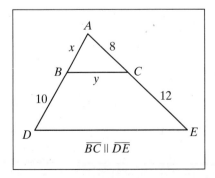

    $\overline{BC} \parallel \overline{DE}$

    b.

    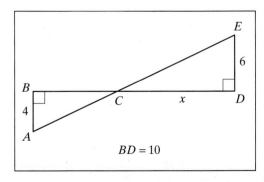

    $BD = 10$

52. The symbol for "is congruent to" is ≅. When stating that two geometrical objects are congruent, the order of the letters used must correspond to congruent parts. For example △ABC ≅ △RST means that segment AB ≅ segment RS and <C ≅ <T. From the information given, determine the correct congruence statement for the following.

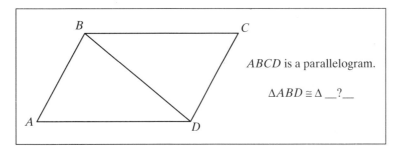

ABCD is a parallelogram.

△ABD ≅ △ __?__

53. Prove that the side opposite the 30° angle in a 30°–60°–90° triangle is half as long as the hypotenuse. (*Hint*: Try making an equilateral triangle out of two congruent triangles.)

54. Find missing side and angle measures in the following pair of congruent triangles. Explain your reasoning.

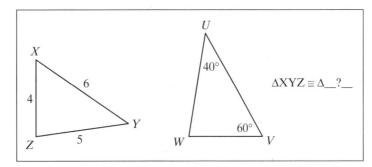

△XYZ ≅ △__?__

55. a. Make a conjecture about the base angles of an isosceles triangle.
    b. Prove your conjecture. (*Hint*: Remember that often it helps to introduce "something new." Draw the isosceles triangle, and mark the congruent sides. Which of the following should you introduce into your picture to help you prove your conjecture: altitude, median, or angle bisector from the apex to the base?)

56. Given △ABC with <BCA and <CDA right angles. Prove that <A is congruent to <DCB. Be sure to justify your statements.

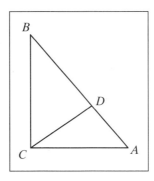

57. Moses said that he was sure that if two angles and the included angle of one triangle are congruent to two angles and the included side of another, respectively, then the two triangles will be congruent (ASA). Rebekah said that AAS could also be used to prove two triangles are congruent. What do you think?

58. Prove the following theorems. (In each case, the results of one can be used to prove the next.)

   a. In any triangle, the line segment formed by connecting the midpoints of any two sides is parallel to the third side and equal to 1/2 the length of the third side.
   b. The quadrilateral formed by joining the midpoints of consecutive sides of any quadrilateral is a parallelogram.
   c. The quadrilateral formed by connecting the midpoints of consecutive sides of a rectangle is a rhombus.

59. Draw an angle, and construct the angle bisector. Prove that any point on the angle bisector is equidistant from the sides of the angle.

60. In any triangle, the line segment connecting the midpoints of any two sides is parallel to the third side. Prove that the smaller triangle formed with this segment is similar to the original triangle.

61. Sketch the following triangles: $\triangle ABC$ is an isosceles right triangle with legs of length 4 cm. $\triangle DEF$ is an isosceles right triangle with legs of length 8 cm.

   a. Find the measures of all angles and sides. Support your answers with logical reasoning and/or use of theorems.
   b. Prove or disprove the statement: These triangles are similar.
   c. Prove or disprove the statement: These triangles are congruent.
   d. Find the area of each triangle.

62. A student claimed that the sum of the measures of the angles in a 10-sided figure is 1,800°. Her proof was to connect a point in the interior with the 10 vertices to form 10 triangles. Thus, $10 \cdot 180° = 1,800°$. What do you think about her proof?

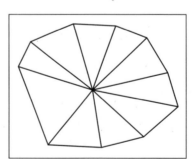

63. Find m∠1 + m∠2 + m∠3 + m∠4 + m∠5 in any five-pointed star. Justify your answer.

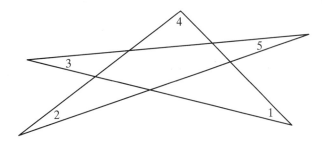

64. Find the sum $a + b + c + d + e + f$ in the figure.

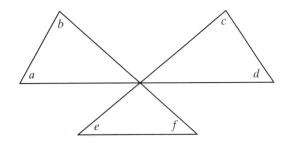

Extending the Activity

65. Define all seven types of quadrilaterals that you considered in Activity 9.9. Remember that a definition must include necessary and sufficient conditions.

66. Using the table you completed in Activity 9.9, consider the angle properties of each quadrilateral, and note relationships between the number of angle properties a quadrilateral has and its position in the tree diagram.

67. Using the table you completed in Activity 9.9, consider the symmetric properties of each quadrilateral, and note relationships between the number of symmetric properties a quadrilateral has and its position in the tree diagram. Count each line of symmetry once, and count rotational symmetry once.

68. a. How many different triangles with a perimeter of 12 units can be made if the length of each side is a whole number? Find as many possibilities as you can, and after each, use toothpicks (1 toothpick=1 unit of length) to verify that you have found numbers that work.
    b. Roll three dice. How likely is it that the three numbers you get can be the lengths of sides of a triangle?
    c. Without actually making any triangles, how can you determine if three numbers can be the lengths of sides of a triangle?

69. a. Suppose your home is 23 miles from Nearsville and 49 miles from Farsville. How far from Nearsville is Farsville?
    b. Find a way to determine how far apart two towns are if you know the distance from each of these two towns to a third town.

70. Suppose you have a stick with a length of 12 feet. If you drop the stick and it breaks into three pieces, how likely is it that the three pieces could be the sides of a triangle? How many different right triangles are possible?

71. a. On a blank sheet of paper, redraw the first $\triangle ABC$ in Activity 9.13 with point P outside the triangle. Instead of marking off twice the length, mark off three times the length. What happens? Do relationships change? If so, how?
    b. Instead of marking off integer multiples of the length, bisect the length. What happens to the new triangle? How do the relationship change?
    c. Redraw the triangle and place point P on the triangle. Follow the steps for constructing a new triangle. How does the new triangle compare with the original triangle?

72. In the discussion in class for Activity 9.14, you may have discussed that when the scale factor is $n/1$, the dilation has area n^2 times the area of the original. For example, when the scale is 2/1, the dilated area is 4 ($= 2^2$) times the original area. When the scale is 1/2, the dilated area is 0.25 ($= 0.5^2$) that of the original. Extend your conjecture to similarity in three dimensions. How many unit cubes (side length 1) would you need to make a (similar) cube whose side length is 2? length 3? length n?

73. Redraw the figure from #4 on Activity 9.15, starting with a rectangle that is not a square. Will the figure drawn by connecting the midpoints be a square? If yes, prove your answer. If not, state what shape it is, and then prove your answer.

74. Draw a convex quadrilateral $ABCD$ where all four sides and angles have different measures. Draw the line segments connecting the midpoints of each side. Label the new quadrilateral $EFGH$. Prove that quadrilateral $EFGH$ is a parallelogram.

Writing/Discussing

75. Make a concept map about geometry. Discuss the links and connections you made in your map.

76. Discuss important factors to consider when writing definitions. Explain why you think these are important, any changes in your understanding of what makes a good definition, and what caused you to change your thinking about definitions.

77. Many times, patterns for projects are shown in magazines. Because it is nearly impossible to show them to the correct measurements, scale drawings appear. How might you turn the scale drawing into an exact-size drawing?

78. Discuss the applications of various theorems to particular problems. How does one decide whether or not the theorem applies? What is the significance of meeting the conditions in the hypothesis before applying the theorem?

79. Discuss the importance of definitions in mathematical structures. How do you come to understand a definition? What role do examples and nonexamples play in your understanding?

80. In the table in Activity 9.16 that lists angle measurements for polygons, there is a column for the sum of the measures of all angles in each of the polygons. Will that total be the same if the polygon is a regular polygon? Why or why not?

81. How does spherical geometry enable you to think differently about geometry in general?

82. Make a second concept map about geometry, and explain the links and connections in the map. Then write a reflection comparing and contrasting your first and second concept maps.

CHAPTER **10** TEN

Measurement

CHAPTER OVERVIEW

Perhaps no part of mathematics is more clearly applicable to everyday life than measurement, the focus of the activities in this chapter. As a consequence of the practicality of measurement in the real world, it is a very important strand in the elementary school mathematics curriculum. In particular, students learn extremely useful measurement skills (e.g., how to use a ruler), concepts (e.g., the concepts of area and perimeter), and key formulas (e.g., $A = l \cdot w$). Just as important is the fact that making measurements can be a source of many, very interesting problems; for example, did you know that two shapes can have the same perimeter but different areas? One natural question that follows from this is "Can two shapes have the same area but different perimeters?" In this chapter, you will investigate these and many other challenging problems.

BIG MATHEMATICAL IDEAS

Problem-solving strategies, conjecturing; verifying; generalizing

NCTM PRINCIPLES & STANDARDS LINKS

Measurement; Problem Solving; Reasoning; Communication; Connections; Representation

Activity		
10.1	Exploring Area and Perimeter
10.2	Perimeter and Area: Is There a Relationship?
10.3	Pick's Formula
10.4	Investigating Length
10.5	Pythagoras and Proof
10.6	Investigating Circles
10.7	Investigating the Circumference to Diameter Ratio
10.8	Investigating the Area of a Circle
10.9	Surface Area and Volume of Rectangular Prisms
10.10	Drawing Rectangular Prisms
10.11	Exploring the Surface Area of Cones
10.12	Investigating the Volumes of Cylinders and Cones
10.13	Geoboard Battleship: Exploring Coordinate Geometry
10.14	Investigating Translations using *Sketchpad*™
10.15	Investigating Rotations using *Sketchpad*™
10.16	Investigating Reflections using *Sketchpad*™
10.17	Tessellations: One Definition
10.18	Tessellations: Another Definition

ACTIVITY 10.1 Exploring Area and Perimeter

FYI Topics in the Student Resource Handbook

10.3 **Perimeter**
10.5 **Measuring Area**

1. The area of a square of side 1 unit length is 1 *square unit*. What is the area of a square of side length 1 inch? 1 mile?

2. Make a drawing *to demonstrate a proof* that the area, A, of a rectangle of length x units and width y units is equal to xy square units (take x and y as whole numbers). Describe your proof below, using a picture, if necessary.

3. Using the result $A = xy$ for A, x, and y as in #2 above and knowledge that a square is a special kind of rectangle, give an explanation for the formula for the area of a square.

4. a. Determine a formula to find the perimeter P of a rectangle of length x units and width y units. Explain why the formula is valid.

   b. Determine a specialized formula to find the perimeter of a square of side s units. Explain why the formula is valid.

5. Devise a strategy for finding the area of the following polygon, and then find the area.

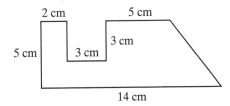

ACTIVITY 10.2 Area and Perimeter: Is There a Relationship?

FYI Topics in the Student Resource Handbook

10.3 Perimeter
10.5 Measuring Area

For each of the following, try to find at least two rectangles that fit the conditions listed. Record your work on graph paper. You are limited to rectangles whose lengths and widths are whole numbers and that can be built using at most 25 square tiles, each tile having area 1 square unit. For a given condition, if no such rectangle exists or only one such example exists, explain why. Is it because you are limited to 25 tiles, or does such a rectangle not exist at all?

1. area between 10 and 16 square units

2. perimeter between 8 and 15 units

3. area numerically less than perimeter

4. area numerically equal to perimeter

5. area numerically greater than perimeter

6. area an even number

7. perimeter an even number

8. area an odd number

9. perimeter an odd number

10. area a square number (but not a square)

11. perimeter a multiple of 4 without length or width being equal to 4 units

12. maximum perimeter

13. maximum area with less than 20 tiles

14. minimum perimeter with an area of 16 square units

15. maximum perimeter with an area of 16 square units

16. perimeter that is neither minimum or maximum for an area of 12 square units

17. area numerically equal to a prime number

18. perimeter numerically equal to a prime number

19. area is numerically one more than perimeter

20. area that is more than perimeter by the same number that designates its width

Looking for a Relationship between Area and Perimeter

1. If two polygons have the same perimeter, can you make a generalization about their areas? Use a geoboard for exploration.

2. If two polygons have the same area, can you make a generalization about their perimeters? Use a geoboard for exploration.

ACTIVITY 10.3 Pick's Formula

Topics in the Student Resource Handbook

10.3 **Perimeter**
10.5 **Measuring Area**

In Activity 10.2, we found that there is not a direct relationship between area and perimeter in general. However, there is a relationship between area and perimeter for polygons on a lattice (like a geoboard) if we consider the interior and boundary points of the lattice.

In 1899, Georg Pick generated a formula for calculating the area, A, of a polygon that can be formed on a geoboard. The formula gives A in terms of just two easily determined numbers, I and B, where

  I = the number of geoboard points in the interior of the polygon, and
  B = the number of geoboard points on the boundary of the polygon.

Your task is to rediscover the relationship between A, I, and B by exploring with geoboard polygons.

To do this effectively, you should consider some systematic way of obtaining your data and make a table to keep track of your results. Although the formula works for *any* geoboard polygon, you should work with those where you can easily determine their area. For example, it is easy to see that polygon #1 below has an area of 8 square units (why?) but that of polygon #2 is not found as easily. However, once you have discovered the formula, you can find the area of polygon #2 in terms of its values for I (6) and B (9).

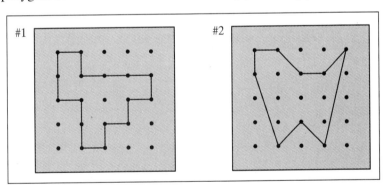

ACTIVITY 10.4 Investigating Length

 FYI Topics in the Student Resource Handbook
10.2 **Linear Measurement**

In each of the figures below, the distance between a dot and any horizontally or vertically adjacent dot is 1 unit. For each square below, find the square's area and side length. Be prepared to describe how you calculated the area.

1.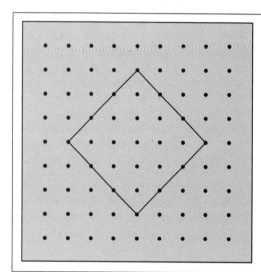

   Area of square = _____

   Length of side = _____

2.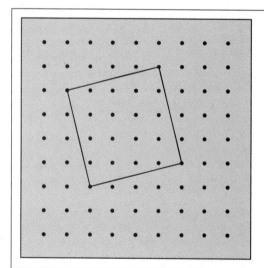

   Area of square = _____

   Length of side = _____

3.

Area of square = _____

Length of side = _____

4. Describe a procedure for finding the area and side length of any tilted square on a geoboard.

5. Can you use the same or a similar method to find the area and side length of this square?

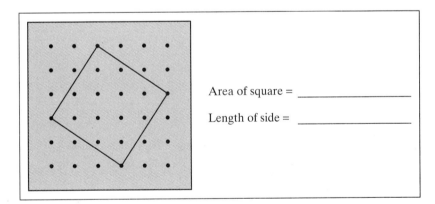

Area of square = _____

Length of side = _____

6. Using the ideas explored above, find the length of the hypotenuse (longest side) of this right triangle.

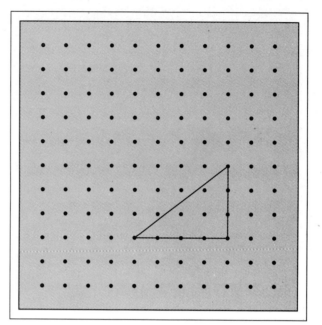

ACTIVITY 10.5 Pythagoras and Proof

Topics in the Student Resource Handbook

10.16 Pythagorean Theorem and Other Triangle Relationships

The Pythagorean theorem is probably the most famous theorem in all of mathematics. The simple equation, $a^2 + b^2 = c^2$, would probably be recognized by anyone who ever took a high school mathematics class, whether they remembered what it meant or not.

Pythagoras was a Greek philosopher who lived in about 500 B.C. Although he is credited with the theorem, a number of different cultures around the world—Arabic, Chinese, Indian, European and probably others—both before and after Pythagoras's time, discovered the right triangle relationship, or at least specific cases to which it applies, independently of one another.

Pythagoras lived and traveled throughout the Mediterranean and Asia Minor, studying and teaching in Egypt, Babylonia, and what are now Italy and Greece. Babylonia was an important center of world commerce, and during his time there, Pythagoras had the opportunity to study with Babylonian, Egyptian, Chinese, and Indian scholars who may have known the theorem. No one knows whether Pythagoras came upon the theorem in his studies or discovered it himself independently. On his return to Greece, Pythagoras established a school that had a lasting influence on the study of mathematics. In violation of the laws of the time, he allowed and encouraged women scholars. His wife and daughters kept the school active long after his death. Pythagoras's proof may be the proof found in Euclid's *Elements* or one of the simpler proofs you will try in class. Pythagoras's claim to fame is that he or people in his school *proved* the theorem, although ironically nobody is sure what Pythagoras's proof of the theorem was. All proofs from Pythagoras's school were attributed to him, even though others may have been responsible for the proofs. It is possible that Pythagoras never proved the theorem that bears his name.

Introduction of the Pythagorean Theorem

Your instructor will first lead the class in a discussion of the meaning of the Pythagorean theorem's familiar result: $a^2 + b^2 = c^2$. A proof of the theorem will then be obtained based on the diagram below. The proof uses ideas of congruence of triangles and decomposition of areas. You should participate in the discussion and write your own proof below the diagram.

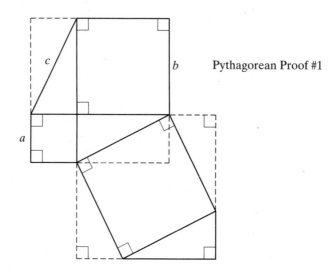

Pythagorean Proof #1

Group Work on Proofs

Although the Pythagorean theorem is so famous, what is not so well known is that there are numerous ways to prove it. You have already discussed one proof of it. On each of the following pages is a diagram that suggests a particular version of proof. You will now get to prove the theorem based on the diagram assigned to your group. Be ready to present your proof to the class at the end of 20 minutes.

Pythagorean Theorem Extensions

The Pythagorean theorem is usually thought of in terms of the additivity of the areas of squares on the two sides of a right triangle (e.g., some of the proofs that were done today used this idea). Instead of drawing only squares, is it possible to draw other figures, such as the ones below? Does the idea of additivity of areas from the Pythagorean theorem hold for these figures? Why or why not?

1.

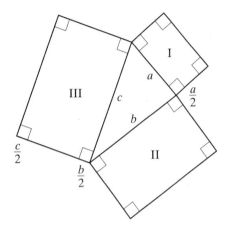

2.

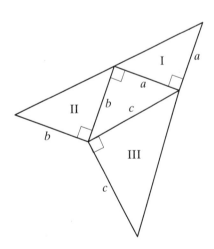

3.

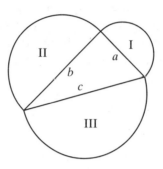

Pythagorean Proof #2 *(The Simplest Dissection)*

The following figure appears in very old texts, and you may have even seen it in decorative tile designs. Nobody knows, but it may have been the proof known to Pythagoras, as it seems to have been known in China and India and as Pythagoras studied with Chinese and Indian scholars while he was in Babylonia.

The Pythagorean Theorem: If a right triangle has legs of length a and b and a hypotenuse of length c, then $a^2 + b^2 = c^2$.

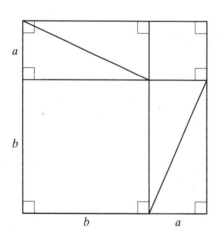
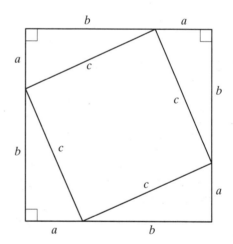

Pythagorean Proof #3 *(Perigal's Proof)*

Many proofs of the Pythagorean theorem have ancient origins but were rediscovered later by people unfamiliar with the older sources. This proof was "discovered" by mathematician Henry Perigal in 1873, but was probably known to the Arabian mathematician Iabit ibn Qorra a thousand years before.

The Pythagorean Theorem: If a right triangle has legs of length a and b and a hypotenuse of length c, then $a^2 + b^2 = c^2$.

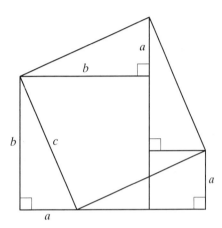

Pythagorean Proof #4 *(Presidential Proof)*

James Abram Garfield (1831–1881), the country's 20th president, discovered a proof of the Pythagorean theorem in 1876 while he was a member of the House of Representatives, five years before he became president of the United States. An interest in mathematics may not be a prerequisite for the presidency, but it had been common at the time. One of Garfield's predecessors, Abraham Lincoln, credited Euclid's *Elements* as being one of the books most influential to his career as a lawyer and a politician, saying he learned from it how to think logically. Garfield's proof of the theorem is illustrated with a relatively simple figure: a trapezoid.

The Pythagorean Theorem: If a right triangle has legs of length a and b and a hypotenuse of length c, then $a^2 + b^2 = c^2$.

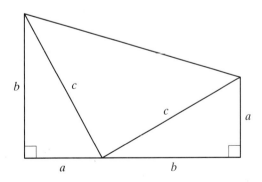

Pythagorean Proof #5 (Behold!)

The 12th-century Hindu scholar, Bhaskara, wrote the single word "Behold!" to accompany his figure demonstrating the Pythagorean theorem. Bhaskara must have felt the figure spoke for itself! Perhaps it does speak for itself, but you can gain a deeper understanding by constructing a proof based on the figure. Incidentally, this figure is also found in an ancient Chinese text, making it another candidate for being a proof known to Pythagoras.

The Pythagorean Theorem: If a right triangle has legs of length a and b and a hypotenuse of length c, then $a^2 + b^2 = c^2$.

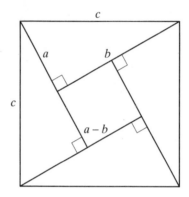

Pythagorean Proof #6

This figure is attributed to Saunderson (1682–1739) but probably came from the 12th-century Hindu mathematician Bhaskara (see Proof #5).

The Pythagorean Theorem: If a right triangle has legs of length a and b and a hypotenuse of length c, then $a^2 + b^2 = c^2$.

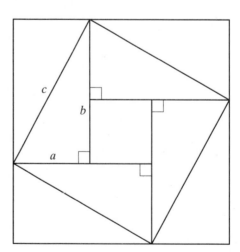

Pythagorean Proof #7 *(Similar Triangle Proof)*

The similar triangle proof has the dual distinction of being the shortest when written out, as well as being the proof most commonly found in geometry books. It is also the basis of a generalization of the Pythagorean theorem that will be discussed at the end of the class period.

The Pythagorean Theorem: If a right triangle has legs of length a and b and a hypotenuse of length c, then $a^2 + b^2 = c^2$.

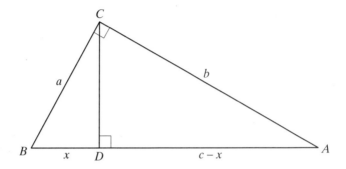

426 Chapter 10 **Measurement**

ACTIVITY 10.6 Investigating Circles

FYI **Topics in the Student Resource Handbook**

  9.11 **Circles**
10.4 **Circumference**

The circle is a geometric shape that is all around you. Some societal uses are wheels, circular gears and machinery parts, potter's wheels, clocks, windmills, and compact discs. Great advances to civilization have been based on the applications of circles and spheres. In this activity, your group will examine and define parts of a circle: the center, radius, diameter, chord, secant, tangent, central angle, and inscribed angle. The following figure is referred to as *circle O* because the center is at O:

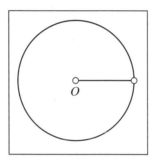

A line segment from the center to a point on the circle is called a *radius*.

Write a definition of each geometric term after discussing the following examples with your group members.

1. Define *chord* _____

_____

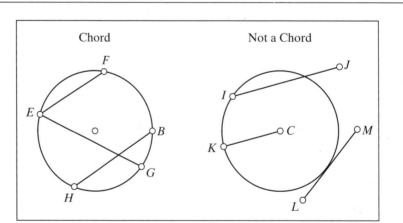

2. Define *diameter* _____
   _____

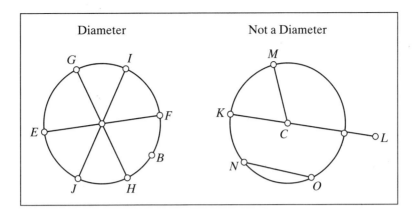

3. Define *secant* _____
   _____

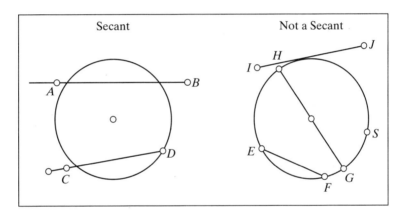

4. Define *tangent* _____
   _____

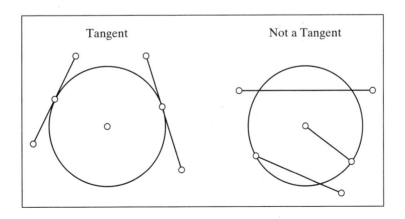

5. Define *inscribed angle* _____

_____

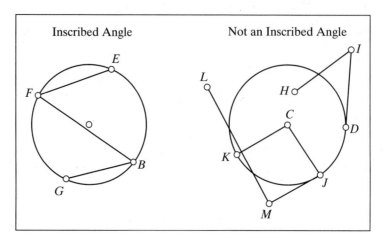

6. Define *central angle* _____

_____

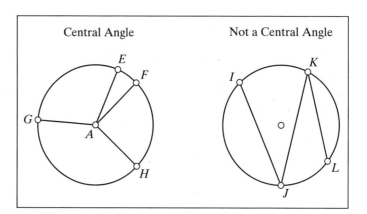

ACTIVITY 10.7 Investigating the Circumference-to-Diameter Ratio

FYI Topics in the Student Resource Handbook

10.4 **Circumference**

In a polygon, the distance around the figure is called the perimeter. In circles, the distance around the figure is called the circumference. In this activity, your group will discover (or perhaps rediscover) the relationship between the diameter and the circumference of every circle. Once you know the relationship, you can measure a circle's diameter and calculate its circumference.

Part 1: Investigation

Materials:
Your instructor will give you two circular objects, string and a yardstick.

Step 1: With the string and the yardstick, measure the circumference and diameter of each round object to the nearest tenth inch.

Step 2: Make a table, and record the circumference (C) and diameter (D) measurements for each round object.

Step 3: Calculate C/D and enter each answer in the corresponding column in your table.

Step 4: Calculate the average of your C/D results.

Record your table below and be sure to include the name of the object.

Part 2: Discussion

Each group will list their average ratio on the chalkboard and be prepared to discuss the similarities and differences between these averages. Give a formula definition of the ratio C/D: _____.

If you solve this formula for C, you get a formula for finding the circumference of a circle in terms of the diameter.

The formula is: _____.

The formula can be rewritten in terms of the radius: _____.

ACTIVITY 10.8 Investigating the Area of a Circle

Topics in the Student Resource Handbook

10.8 Area of Regular Polygons and Circles

In this activity, your group will develop a formula for calculating the area of any circle. Each group member will perform a separate part of the investigation.

Part 1: Investigation

Materials:
Each group member will need a compass, a pair of scissors, a sheet of construction paper, and a glue stick.

Step 1: With your compass, make a circle with a radius of approximately 3 inches. Cut out the circular region.

Step 2: Fold the region in half. Fold it in half a second time. Fold it in half a third time. Fold it in half one last time.

Step 3: Unfold the circular region and cut it along the folds into 16 wedges.

Step 4: Glue onto the construction paper a row arrangement of the 16 wedges, alternating tips up and down to form a shape that resembles a parallelogram.

1. What would the shape resemble if you cut more wedges?

2. Could you say the area of this new shape is the same as the area of the original circle? Why?

3. Sketch your wedge shape below and label the dimensions. Give a reasonable algebraic and written argument for your group's conjecture of the area of circle.

Part 2: Application

Discuss the following problems within your group. Decide how to solve them, and provide a detailed solution based on that agreement. Include diagrams and reasons for steps you take and why certain formula(s) are necessary.

1. A small college TV station can broadcast its programming a radius of 60 km. How many square kilometers of viewing audience does the radio station have? Use 3.14 for π. Explain why your answer is in terms of square kilometers.

2. The strength of a bone or a muscle is proportional to its cross-sectional area. If a cross-section of one muscle is a circular region with a radius of 3 cm and a second identical type of muscle has a cross-section that is a circular region with a radius of 6 cm, how many times stronger is the second muscle?

ACTIVITY 10.9 Surface Area and Volume of Rectangular Prisms

FYI Topics in the Student Resource Handbook
10.9 Surface Area of Prisms and Cylinders

1. A cube is a familiar polyhedron.

   a. Describe some common objects that have the shape of a cube.

   b. Describe as many properties of a cube as possible.

   c. Define a *cube*.

   d. What is the volume of a cube with edge 1 cm?

2. *A rectangular prism is a polyhedron with all its faces rectangular.* Discuss why this definition forces a rectangular prism to have exactly six faces. [*Hint:* Think about what is an interior angle of a face and how many such faces can meet at a vertex so as to form a solid.]

3. Using cubic blocks, demonstrate a proof that the volume V of a rectangular prism of length x, width y and height z units is given by $V = xyz$ cubic units. Assume that a cubic block has volume 1 cubic unit and that $x, y,$ and z are whole numbers. Describe your proof.

4. The surface area of a solid is the sum of the area of all its faces. What is the surface area, S, of a rectangular prism with length x, width y, and height z units? Explain your answer.

5. Use your result in question #4 and the knowledge that a cube is a special kind of rectangular prism to determine the surface area, S, of a cube of edge s units.

ACTIVITY 10.10 Drawing Rectangular Prisms

FYI Topics in the Student Resource Handbook
10.9 Surface Area of Prisms and Cylinders

1. A rectangular prism is a three-dimensional object. To draw a representation of a three-dimensional object on two-dimensional paper is usually not easy, but doing so for rectangular prisms (especially cubes) is not too difficult.

   To draw a cube, first draw a square to represent its front face and a slightly smaller square behind it to represent the back face.

   Now, join the corresponding vertices of the two squares, using straight line segments.

   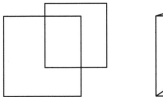 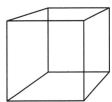

   a. We indicate which edges of the cube cannot be seen by the viewer by using dotted lines. In each of the following pictures of cubes, describe, using one from each of the following two groups of qualifiers, where the viewer is in relation to the cube: above/below and left of/right of/in front of.

   (i) 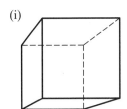 (ii) 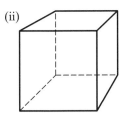 (iii) 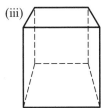 (iv)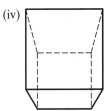

   b. Draw a picture of a cube of edge length 5 cm, viewed from slightly above and to the left.

2. The drawings of cubes in #1 capture the impression that objects further away from the viewer (like the back face of the cube) appear smaller. This is a characteristic of *perspective drawing*, which tries to convey a feeling of depth on a two-dimensional plane.

   The drawings in #1 were all examples of perspective drawing with *one-point perspective*, where a face of the cube was taken parallel to the plane of the paper so that all lines, which cannot be contained on a plane parallel to that of the paper, *meet in a single point* when extended.

   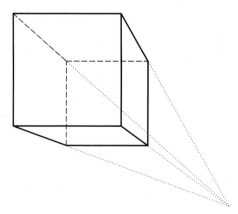

   In this drawing, the single point is below and to the right of the cube. Where is the viewer in relation to the cube?

3. An example of a *two-point perspective drawing* of a cube is shown below.

   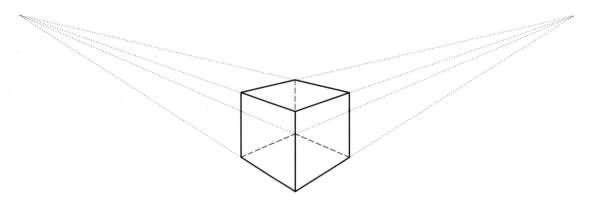

   a. Is a face of the cube parallel to the plane of the paper?

b. Which part of the cube is closest to the viewer?

4. A technique similar to that used for drawing cubes in question #1 can be used to draw any rectangular prism. Use it to draw a rectangular prism of dimension 5 cm × 4 cm × 4 cm so that:

   a. the square faces are parallel to the plane of the paper and the rectangular prism is viewed from slightly below and to the right.

   b. a pair of nonsquare faces is parallel to the plane of the paper and the rectangular prism is viewed from slightly above and directly in front.

440 Chapter 10 Measurement

ACTIVITY 10.11 Exploring the Surface Area of Cones

FYI Topics in the Student Resource Handbook

10.10 Surface Area of Pyramids, Cones, and Spheres

A cone has three important linear features. They are the *radius*, the *height*, and the *slant height*.

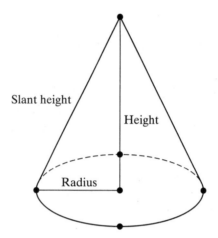

Cut out the circle in the Appendix on the page indicated by your instructor. Cut through to the center of the 8-cm radius line. By placing the cut line on top of the lettered surface of the circle and moving the edge of the cut line to one of the lettered positions, the lateral surface of a cone is formed. Use a paper clip to hold it in place. Slowly move the cut edge from letter A to B to C, and so on. The height of the cone will increase.

1. Does the radius increase, decrease, or remain the same?

2. How does the slant height change?

3. What is the longest radius possible?

4. What is the greatest height possible?

This time open the model of the cone from a tightly overlapped position by moving the edge back toward the letters C, B, and A. The height of the cone will decrease.

5. At what height is the lateral area the greatest?

6. At what height is the area of the base the least?

7. a. As the height of the cone decreases, the lateral area increases. What happens to the area of the base?

Activity 10.11 *Exploring the Surface Area of Cones*

b. What happens to the total surface area?

8. a. As the cone is opened, the area of the base approaches the area of the original 8-cm circle. Does the lateral area approach the same value?

   b. Does the total area approach the same value?

9. Make the cone formed by placing the cut line on D. Find the height of the cone.

10. Using an 8 1/2-by-11-in. piece of paper, what is the maximum height you could have when making a cone?

11. If the circumference of the cone is 3 cm and the lateral surface area is 25 square cm, can you find the height of the cone? If yes, do it. If no, why not?

12. Given the circumference and the slant height of any cone, can you find the lateral surface area of the cone? Explain.

13. Is it possible for two cones to have the same lateral area and different heights? Explain.

14. Is it possible for two cones to have the same base area, different heights, and the same lateral area?

15. If you know the lateral area, can you predict the height? If yes, explain your answer. If no, give an example.

ACTIVITY 10.12 Investigating the Volumes of Cylinders and Cones

FYI Topics in the Student Resource Handbook
10.12 Volume of Prisms and Cylinders
10.13 Volume of Pyramids, Cones, and Spheres

In this investigation, your group will build a cylinder-cone pair with congruent bases and the same height. The following definitions will help familiarize you with these objects:

*A **cylinder** is a solid composed of two congruent circles in parallel planes, their interiors, and all the line segments parallel to the axis with endpoints on the two circles.*

The circles and their interiors are the **bases**. The **radius (*r*)** of the cylinder is the radius of a base. The **altitude** of a cylinder is a perpendicular segment from the plane of one base to the plane of the other. The **height (*h*)** of a cylinder is the length of an altitude.

If the axis of a cylinder is perpendicular to the bases, then the cylinder is a **right cylinder.**

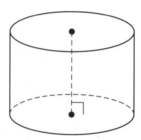

Use your knowledge of the area of a circle and that volume is a three-dimensional measure to give a formula for the volume of the "disk," shown below, whose height is 1 unit:

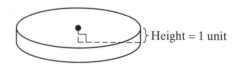

Now, generalize this formula for the volume of any cylinder: _____

*A **cone** is a solid composed of a circle, its interior, a given point not on the plane of the circle, and all the segments from the point to the circle.*

The circle and its interior make up the **base** of the cone. The **radius** of the cone is the radius of the base. The point is the **vertex** of the cone. The **altitude** of a cone is the perpendicular segment from the vertex to the plane of the base. The **height** of a cone is the length of the altitude.

If the line segment connecting the vertex of a cone with the center of its base is perpendicular to the base, then it is a **right cone**. In this investigation, you will discover the formula for the volume of a cone.

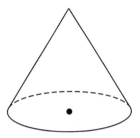

Build a Cylinder-Cone Pair

Your instructor will provide your group with the following needed materials:

- a large, clean, empty can
- a manila folder
- scissors and tape
- a paper bag filled with a quantity of rice sufficient to fill the can

Step 1: Force a manila folder rolled into a cone into the can until the tip firmly touches the bottom of the can. With the help of another group member, adjust the open end of the cone so that it is tight against the rim of the can. Tape the cone securely.

Step 2: Mark a circle on the cone where the rim of the can touches the cone. Cut off the excess manila folder along the circle.

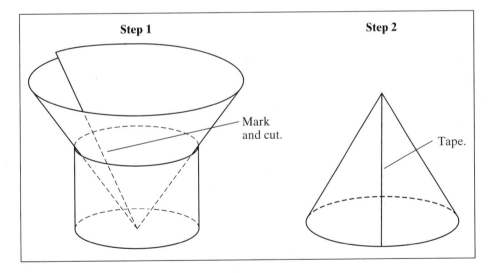

You now have a cylinder and cone with congruent bases and the same height.

Step 3: Discuss within your group how you might discover the formula for the volume of the cone using the rice. Determine the best procedure to use and then carry it out. [*Hint*: Use the rice!]

Explain and describe your procedure and conjecture. What did you discover to be the formula for the volume of your cone?

Maximizing the Volume of a Cone

This investigation will allow your group to explore what happens to the volume of a cone when various dimensions change. You will need to cut out the circle in the Appendix on the page indicated by your instructor and use the centimeter strip for measuring. Obtain a paper clip from your instructor.

Make the cone tighter by moving the edge from letter A to B to C, and so on. The height will increase, and the area of the base will decrease.

1. Describe in your own words what happens to the volume.

2. Discuss within your group what position forms the cone with the greatest volume. Take your model and set it at that position. Paper clip it in place.

    Letter _____

3. Measure to the nearest 0.5 cm the height of the cone in that position. Find the radius.

    Radius _____

4. Carefully measure to the nearest 0.5 cm the height of the cone in that same position.

   Height _____

5. Find the volume for the cone at this position. Use 3.14 for π, and show work.

   Volume _____

6. Plot your radius and volume on the same axes below. Then, plot the volume for at least four different radius values.

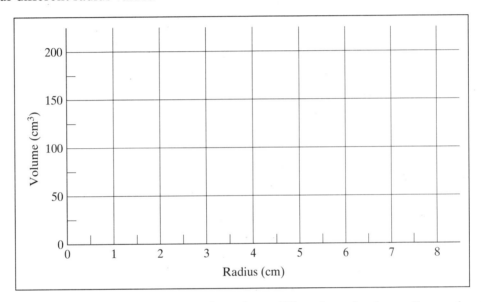

Explain how dimension changes affect the volume. What does the data tell you about maximum volume?

ACTIVITY 10.13 Geoboard Battleship: Exploring Coordinate Geometry

FYI Topics in the Student Resource Handbook

8.3 Cartesian Coordinate System

Form two teams within your group. Each team gets a 5 × 5 geoboard and three rubber bands. Team A places rubber bands on their geoboard, which is hidden from Team B, to represent their "fleet" that consists of:

- A gunboat, represented by a square of side 1
- A destroyer, represented by a square of side 2
- A submarine, represented by a 1 × 2 rectangle

The sides of all three rectangles must be parallel to an edge of the geoboard, and none of the ships may touch the others.

Team B tries to sink the fleet by trying to guess the exact location of each ship. A geoboard peg is specified by two coordinates (column #, row #). The far left column is column 1, and the bottom row is row 1.

Team B should call out "shots" by specifying their coordinates (always using whole numbers). Team A must specify whether the shot is a "miss" (in the water) or a "hit." If the shot is a hit, team A should specify whether the shot is *within* the ship, on a *vertex*, or on the *perimeter* (excluding the vertices). At no time should team A reveal *which* ship is hit. Team B should keep an organized record of each shot: whether a "hit" or a "miss" and, if a hit, what kind.

The game ends (and teams switch roles) when team B correctly identifies the precise location of each ship in the fleet. The goal is to discover the location of each ship in the fewest number of shots possible. When team B is sure that they have correctly identified all the locations, they should mark those locations on their own geoboard and compare with team A.

Once both teams have played both roles, answer the following questions:

1. If a shot fell *within* a ship, which ship got hit?

2. When team A places its ships, how many choices do they have for the placement of the gunboat? Of the destroyer? Of the submarine?

3. If the destroyer has been placed, how many choices are there for placing the gunboat?

4. When team B makes its first shot, what is the probability that it lands in the water?

5. In a certain game, team B made their first shot and immediately knew the precise location of the destroyer. Which guesses could have led so quickly to this discovery?

6. Describe a situation, including the coordinates of the fleet and of the guesses, in which team B could discover the location of all three ships with only two shots.

7. Suppose this game were played with the same ships and the same rules but on a 6 × 6 geoboard. In a "best-case" scenario, what is the fewest number of shots needed to find the fleet?

ACTIVITY 10.14 Investigating Translations using the Geometer's Sketchpad™ (Optional—Requires use of Geometer's Sketchpad™)

FYI Topics in the Student Resource Handbook

10.21 Translations

There are many ways to explore translations with the *Geometer's Sketchpad™* In this activity, however, you will work with only one of these, *Translation by a Marked Vector*. A vector represents part of a line that has length and direction. You will investigate and then define a translation. [Reminder: Before choosing a menu item for an object, you must "select" the object with the arrow tool. For more than one object, hold the shift key while selecting the objects.]

Sketch

Step 1: Open a New Sketch, and use the point tool to create the vertices of a polygon (e.g., a letter or other shape). *Construct the polygon interior.*

Step 2: Create a segment less than 2 inches long and not far from your polygon. Label the endpoints *A* and *B*, using the label tool. Select the two endpoints of your segment, noting the order of selection.

Step 3: Under the *Transform* menu, select *Mark Vector*, noting the name of the vector.

Step 4: Select your polygon; then, under the *Transform* menu, select *Mark Vector*.

Step 5: When the Translate window appears, choose *By Marked Vector*. What happens?

Example:

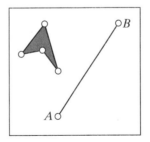

Investigation

1. Using the arrow tool, investigate what happens when you move any point of the pre-image. What effect does this have on the image?

2. Move one end of your segment that defines the translation vector. What happens to the translated image of your polygon?

3. Measure the length of the sides of your polygon and the area (labels may be useful here). What are similarities or differences you observe between moving point(s) of the pre-image and endpoints of the vector?

4. Discuss with your classmates, and determine the effect of a translation on an object. State as many conjectures as you can. Give an informal definition of a translation.

Extension: Construct a parallelogram using *Transformations by Marked Vector*.

Sketch

Step 1: On a New Sketch, create two translation vectors that share a common point.

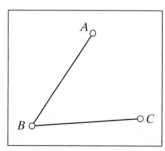

Step 2: Select points *B* and *A* (in that order), and choose *Mark Vector* under the *Transform* menu. Select segment *BC*, and translate it by marked vector *BA*.

Step 3: Select points *B* and *C* (in that order), and choose *Mark Vector* under the *Transform* menu. Select segment *BA*, and translate it by marked vector *BC*.

5. What happens when you change the position of the points *A*, *B*, and *C*?

6. Does your definition hold for the parallelogram? Why or why not?

452 Chapter 10 *Measurement*

ACTIVITY 10.15 Investigating Rotations using the Geometer's Sketchpad™ *(Optional—Requires use of Geometer's Sketchpad™)*

FYI Topics in the Student Resource Handbook
10.22 **Rotations**

There are several ways to directly rotate objects using this software. In this activity, you will be rotating your selection by a fixed angle around a point marked as center. You will be able to examine effects of this transformation and to define it informally.

Sketch

Step 1: Open a New Sketch, and under *Graph* menu, choose *Show Grid*. This background graph will help you to analyze the rotations.

Step 2: In the first quadrant, construct a triangle or other polygon of your choice.

Step 3: Select the origin, and choose *Mark Center* in the *Transform* menu. The object will now rotate around this center point.

Step 4: Select the polygon interior, and choose *Rotate* under the *Transform* menu. Choose 90 degrees for your first investigation.

Step 5: Give your new image a different color or shade.

Step 6: Keep Rotating: Select the new image each time, and choose *Rotate* under the *Transform* menu.

Investigation

1. Investigate and state what happens when you move any point of the original polygon.

2. Compare measurements of the original polygon with those of the rotated images (labels may be useful here). What are some similarities or differences between the pre-image and images? Note your measurements. Obtain a printout of your sketch.

3. Open a New Sketch, and repeat the Steps twice. Each time, choose angles different from 90 degrees. (*Note*: Sketchpad™ cannot evaluate angles greater than 180 degrees.) Compare your findings with your investigation of the 90 degree rotation. Obtain a printout of each of the above sketches.

4. Discuss with your classmates, and determine the effect of a rotation on an object. State as many conjectures as you can. Give an informal definition of a rotation.

5. State any similarities between your investigation of translations and rotations.

454 Chapter 10 Measurement

ACTIVITY 10.16 Investigating Reflections using the Geometer's Sketchpad™ (Optional–Requires use of Geometer's Sketchpad™)

FYI Topics in the Student Resource Handbook

10.23 Geometric Reflections

Sketchpad™ will reflect a selected object across a line, a ray, or an axis, marked as a mirror. In this activity, you will investigate the images of objects reflected across segments using *Mark Mirror*. You will be able to define informally the effects of this transformation.

Sketch

Step 1: Open a New Sketch, and create a triangle on the left side of the screen. Use the point tool; then, select all three points at once, using the shift key. Choose *Segment* under the *Construct* menu so that the three sides of the triangle appear. Label the vertices.

Step 2: Using the segment tool, draw a vertical segment in the middle of the screen (holding the shift key while drawing the segment will ensure a straight segment). While the segment is still selected, choose *Mark Mirror* from the *Transform* menu.

(If you don't see this as an available option, then you either have something else selected as well as your segment, or your segment is not selected. Select the vertical segment, and try again.)

Step 3: Select your whole triangle by dragging the mouse with the button pressed from top left to bottom middle of your screen (a dotted rectangle should appear—make sure your triangle is completely contained in the rectangle). On releasing the mouse button, all objects that were in the rectangle should be selected.

Step 4: With your triangle selected, choose *Reflect* from the *Transform* menu. A mirror image of your triangle should appear on the other side of your "mirror" segment.

Investigation

1. Investigate what happens when you move points of the original triangle. What happens to the mirror image? Move your triangle around; make it bigger, smaller; move it closer to the mirror, further away.

Activity 10.16 *Investigating Reflections using the* Geometer's Sketchpad™

2. Compare measurements of the original triangle with the image (labels may be useful here). What are the similarities or differences? Obtain a printout of your sketch.

3. Create a horizontal mirror perpendicular to the vertical mirror, and reflect both triangles across it. Explore and predict the motions of the three image triangles when you change your original triangle. How are the images alike? Different? Obtain a printout of your sketch.

4. Explore what happens when you change the relative positions of the two mirror segments. How do the images change? Stay the same?

5. Discuss with your partner, and determine the effect of a reflection on an object. State as many conjectures as you can. Give an informal definition of a reflection.

ACTIVITY 10.17 Tessellations: One Definition

FYI Topics in the Student Resource Handbook
10.18 Tessellations
10.19 Tessellations of Regular Polygons

1. A *polygonal region* is a polygon together with its interior. Some polygonal regions can be used to tile a plane. For example, many public places have floors tiled with squares, hexagons, or rectangles. Some people use combinations of shapes—like octagons and squares or hexagons and triangles—to tile various rooms in the house. Notice that the only region the polygons have in common is their sides and that there are no empty spaces between the polygonal regions. Also notice that some of these arrangements have the polygonal regions placed so that all the corners meet at a vertex, while other arrangements have corners that do not meet at the vertex. For example, the rows of bricks used on the outside of a house are alternated; the corners do not all meet at the same place. When the corners do not necessarily meet at a corner, they are called *tilings*. When the polygonal regions are arranged so that all corners meet at a vertex, they are called *tessellations*. Identify the tessellations and tilings below.

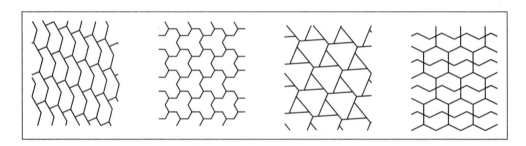

2. When the polygonal regions are arranged so that the corners always meet at a vertex, we can describe the tessellation by listing its *vertex arrangement*. We write vertex arrangements by listing the number of sides in each polygonal region that meets other regions at that vertex. Although it is not important whether we move around the vertex in a clockwise or a counterclockwise direction, it is important to move in a circle rather than randomly listing the number of sides. Some tessellations and their vertex arrangements are shown below.

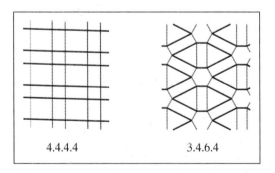

3. A tessellation is a *regular tessellation* if it is constructed of regular polygons. Some arrangements of polygonal regions that tessellate the plane use only one regular polygon; that is, if the arrangement uses squares, no other shapes are used with the squares. Your instructor will give you some regular shapes so that you have at least 10 equilateral triangles, squares, regular pentagons, regular hexagons, regular heptagons, and reg-

ular octagons in your group like the ones on the next page. Try tessellating the plane with each of these figures. Find the regular polygonal regions that will tessellate the plane. List them below. Be prepared to give an analytical (or algebraic) reason why these are the only shapes that will work.

4. A tessellation is a *semiregular tessellation* if it is made with regular polygons such that each vertex is surrounded by the same arrangement of polygons. A few semiregular tessellations are shown below.

   a. How many such arrangements are possible? Be able to justify your answer. [*Hint*: There are more than five and fewer than 10 arrangements.]

   b. List the vertex arrangement for each of the patterns. [*Hint*: You may want to recall the angle measurements of regular polygons from Activity 9.16. It will help you if you are systematic in keeping track of your work.]

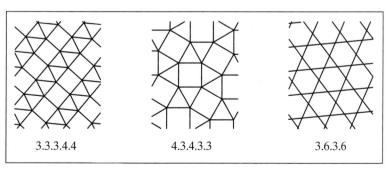

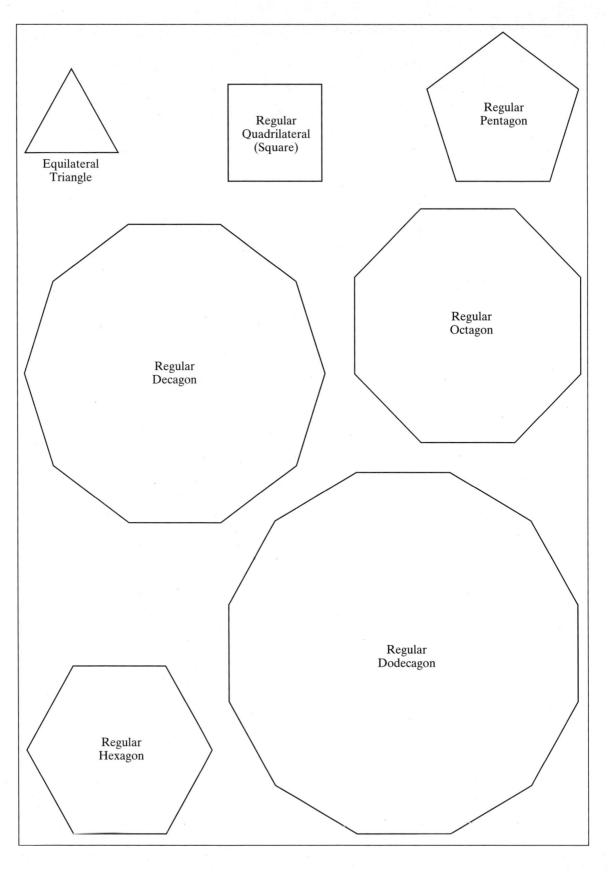

ACTIVITY 10.18 Tessellations: Another Definition

 Topics in the Student Resource Handbook

10.18 Tessellations
10.19 Tessellations of Regular Polygons

1. In Activity 10.17, we began our discussion about tessellations by defining a regular tessellation to be constructed entirely of one regular polygon. If we change the definition of tessellation to include all tilings of the plane, must we change the definition of regular tessellation to have the identical pictures of regular tessellations that we had in Activity 10.17? If you think we need to change the definition, write your new definition below. Test all possible ways of interpreting your definition by drawing all the pictures that fit your new definition. If you think the definition is still okay, test all possible ways to interpret it by drawing pictures that fit the old definition.

2. Does this new definition of tessellation change the definition of *semiregular tessellations*? Explain your answer.

3. The *dual of a tessellation* is the tessellation obtained by connecting the centers of the polygons in the original tessellation that share a common side. The dual of a tessellation of regular triangles is the tessellation of regular hexagons. Find the dual of the other regular tessellations, using either definition of regular tessellation.

4. Will the dual change depending on the definition you use? Explain.

5. Is there any predictable relationship between a tessellation and its dual?

Things to Know from Chapter 10

Words to Know

- altitude
- base
- chord
- circumference
- cube
- diameter
- height
- lateral area
- perimeter
- polygon
- polyhedron
- Pythagorean theorem
- rectangular prism
- regular tessellation
- rotation
- secant
- slant height
- surface area
- tangent
- tiling
- translation
- vertex arrangement

- area
- central angle
- circle
- cone
- cylinder
- dual of a tessellation
- inscribed angle
- line symmetry
- pi (π)
- polygonal region
- prism
- radius
- reflection
- right [cone, cylinder, prism]
- rotational symmetry
- semiregular tessellation
- solid
- symmetry
- tessellation
- transformation
- vertex
- volume

Concepts to Know

- what length is
- what area is
- what surface area is
- what volume is
- what the relationship, if any, between perimeter and area of a polygon is
- what the Pythagorean theorem tells us
- how to prove something
- what a circle and its relationship to a polygon is
- what ρ represents

- the relationships among various polygons and polyhedra
- how translations, rotations, and reflections transform figures
- what tessellations and tilings are
- why some regular polygons tessellate the plane and some do not
- why some polygons always tessellate the plane
- how different definitions may change the structure of some ideas

Procedures to Know

- finding length, area, perimeter/circumference, surface area, volume of various figures

- generating rectangles to meet specific criteria

462 Chapter 10 *Measurement*

- finding the length of a side in a right triangle when given the other two sides
- proving the Pythagorean relationship
- identifying various parts of two- and three-dimensional figures
- drawing rectangular prisms from different views
- translating, rotating, reflecting figures
- making tessellations
- identifying vertex arrangements for tessellations

Exercises & More Problems

Exercises

1. The length of a rectangle is 20 in. The width is 10 in.
   a. What is the area?
   b. What is the perimeter?

2. The area of a rectangle is 24 cm². One dimension is 6 cm. What is the perimeter? Show how you found this answer.

3. The area of a rectangular parking lot is 24 yd². Find all the possible whole-number dimensions of the parking lot in yards.

Find the area of the polygons in #4 and #5 in square feet. The scale is 1 unit=1 ft.

4.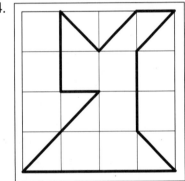

   A = _____

5.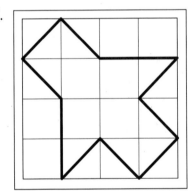

   A = _____

6. In a right triangle, if the hypotenuse is 11 cm and the length of one side is 4 cm, then what is approximately the length of the third side?

7. What is the area of a triangle with an altitude measuring from side AC to vertex B equaling 13 and the length of that same side AC equaling 7?

8. If the area of a rhombus measured 20 cm² and the height measured 4 cm, then what would the length of each side be?

9. If the area of a trapezoid is 75 cm² and the lengths of the two parallel sides are 7 cm and 3 cm, what is the height of the trapezoid?

10. If the ratio between the sides of two squares is 5:7, what is the ratio of the areas? Explain your result.

11. The ratio of the lengths of the corresponding sides of two similar rectangles is 2:1. What is the ratio of their areas?

12. Draw three rectangles that have the same perimeter but different areas.

13. Determine whether the following can be lengths of the sides of a right triangle:
    a. 4, 7, $\sqrt{69}$ b. 6, 8, 10 c. 4, 5, 6

14. If the radius of a ball is 6 cm, what is the surface area of the ball? What is the volume of the ball?

15. What is the surface area of a cylindrical metal food container with diameter 5 cm and height 10 cm, where the top cover is removed?

16. If the dimensions of a rectangular prism are 2 cm × 4 cm × 8 cm, what would be the dimensions of a cube that had the same volume?

17. Which of the following is a better bargain: an orange with radius 4 cm that costs 18 cents, or an orange with radius 5.5 cm that costs 37 cents? Explain.

18. A round waterbed mattress measures 7 feet in diameter by 8 inches thick. How many gallons of water are in this waterbed? Use 22/7 for π.

19. What solid would result if you spun each two-dimensional figure 1–7 about the axis indicated? Match each two-dimensional figure on the top row with a solid of revolution on the bottom row.

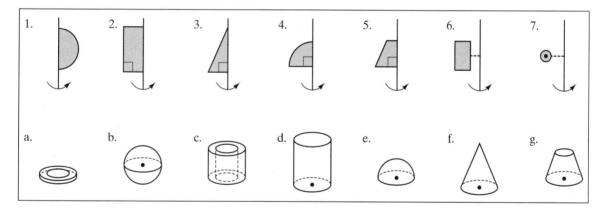

20. Match each three-dimensional figure sliced by a plane with its cross-section.

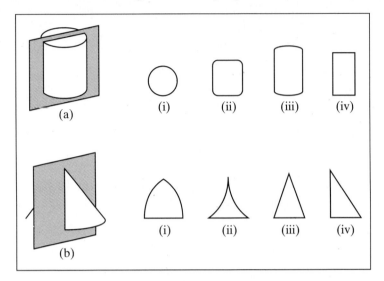

21. Construct a polygon of your choice, and then perform the following rotations. Determine which rotations ended in the same result:

    a. 90° b. 180° c. −270° d. −90°

22. If a triangle with the coordinates A (3,6), B (9,2), and C (−1, −2) is translated three units to the right and two units up, what would the new coordinates be?

23. What words produce the same word when reflected through a horizontal line (e.g., HI)? What about a vertical line (e.g., MOM)?

24. Which of the following properties are not affected by a size transformation? Explain:

    a. the angle measurements   b. lengths of sides   c. proportionality of sides

25. A geometric figure has a *line of symmetry,* 1, if it is its own image under a reflection in 1. Draw all the lines of symmetry for the figures below.

    a.

    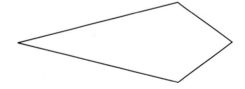

    b.

c.

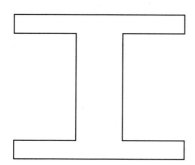

26. Half of a miniature golf hole is drawn below. Complete the figure so that the line shown is a line of symmetry.

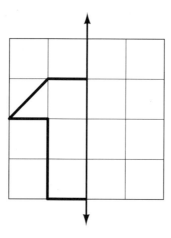

27. Draw the following polygons, and determine how many lines of symmetry exist:

    a. square   b. rectangle   c. equilateral triangle   d. rhombus

28. A figure has *rotational symmetry* when the traced figure can be rotated less than 360° about some point so that it matches the original figure. Determine what rotational symmetries the figures in #27 have.

Critical Thinking

29. If a rectangular field is 80 m × 70 m, how can you determine the dimensions of a square field that has the same area? If you were fencing in this area, how would you determine the lengths of this new square field?

30. How could you represent on a graph the relationships between the length and width of all rectangles with perimeters of 16 cm?

31. Find the areas of the following polygons on the simulated geoboard.

   a. Area of A = _____
   b. Area of B = _____

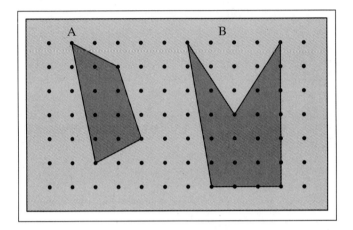

32. Find the perimeter and area of the geoboard figure below:

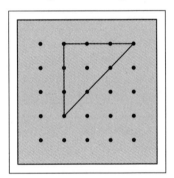

33. Around the world in 80 days? If the diameter of Earth is 8,000 miles, calculate your speed in miles per hour if you were to take 80 days to circumnavigate Earth about the equator. Use 3.14 for π.

34. The Pizza Palace is known throughout the city for its delicious pizza with extrathick crust around the edge. Their small pizza has a 6-inch radius and sells for $9.75. The medium size sells for $12.00 and is a savory 8 inches in radius. The large is a hefty, mouth-watering 20 inches in diameter and sells for $16.50. Because the edge is the thickest part of a Palace pizza, calculate which pizza gives the most edge per dollar.

35. Give as much information as you can about each of the following figures:

    a.
    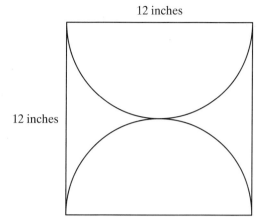

    b.
    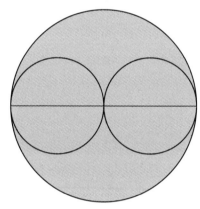

    The diameter of the small circles is 6 cm

36. Explain why rotations sometimes, always, or never result in a similar transformation as a reflection.

37. Do all quadrilaterals tessellate the plane? Explain.

38. a. Which regular polygons tessellate the plane by themselves?
    b. What does the vertex arrangement 3.6.3.6 mean?
    c. Draw (a sufficient portion of) the semiregular tessellation with the vertex arrangement 3.6.3.6.
    d. Why does the vertex arrangement 3.6.3.6 form a tessellation?

39. Adam wants to cover his bathroom walls completely with tiles in the shape of some regular polygon (all tiles used must be the same). List his choices, and make a sketch showing what each would look like.

40. Could 4.4.6.8 describe a vertex point of a semiregular tessellation? Explain.

41. It costs $50 for paint to cover a chalkboard at school. How much would it cost for paint to cover a board half as long and half as wide?

42. A stack of rectangular paper sheets weighs 3 lb. What will be the weight of a stack of sheets of the same paper quality that is three times as high with sheets three times as long and three times as wide?

43. Suppose a wire is stretched tightly around Earth. The radius of Earth is approximately 6,400 km. If the wire is cut, its circumference increased by 20 m, and the wire is then placed back around Earth so that the wire is the same distance from Earth at every point, could you walk under the wire? Justify your answer.

44. Explain two methods for approximating the area of a circle, without using the formula for the area.

45. Draw and label the units of a polygon that has a larger numerical value for its perimeter than it does for its area. Do the same for a polygon that has a larger numerical value for its area than it does for its perimeter.

46. A circular flower bed is 6 m in diameter and has a circular sidewalk around it 1 m wide. Find the area of the sidewalk in square meters.

47. If the ratio of the lengths of the corresponding sides in two similar triangles is 3:1, what is the ratio of their areas?

48. If two trapezoids are similar and the ratio of their corresponding parts is 3:1, what is the ratio of their areas? Explain.

49. a. If the length and the width of a parallelogram are each doubled, is the area of the parallelogram doubled? Explain.
    b. If the base and the height of a triangle are each doubled, what happens to the area of the triangle?

50. Congruent circles are cut out of a rectangular piece of tin, as shown below, to make lids. Find what percent of the tin is wasted.

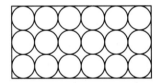

51. Suppose the ratio of the radii of two circles is 2:1.
    a. What is the ratio of the areas of the two circles? Explain.
    b. What is the ratio of the circumference of the two circles? Explain.
    c. If two cylinders of equal height have the circles above as bases, what is the ratio of the volumes of the two cylinders? Explain.
    d. If two cones of equal height have the circles above as bases, what is the ratio of the volumes of the two cones? Explain.

52. Does doubling the radius of a cylinder increase the volume more than doubling the height? Explain.

53. If you were asked to determine the area of a shape that does not have a specific formula, how would you go about determining the area? Draw an example of such an irregular figure, and determine the area of it.

Extending the Activity

54. Find the area of the octagonal-shaped miniature golf hole whose dimensions are given below. Show and explain all your work.

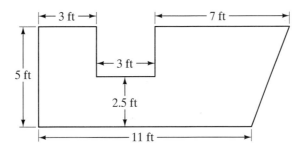

55. Find the area of the miniature golf hole shown below. Explain how you obtained your answer.

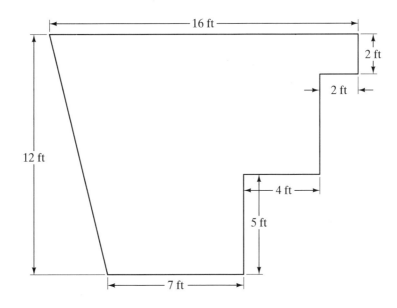

56. If two polygons on a geoboard have the same number of boundary pins, will they also have the same perimeter? Explain.

57. How many noncongruent polygons on a geoboard can be constructed that have five boundary pins and two interior pins? Explain.

58. If two polygons on a geoboard have five boundary pins and two interior pins, do they also have the same perimeter? Do they have the same area? Explain.

59. Draw a proof without words to show that $1 + 2 + 3 + 4 + \cdots + n = n(n + 1)/2$ Proving this is the same as proving $2(1 + 2 + 3 + 4 + \cdots + n) = n(n + 1)$. Hint: The Cereal Box and Patio Tile problems in Activity 1.3 may help.

Write the explanation for each of these "proof without words" in 60–62.

60. Area of a parallelogram is $b \cdot h$

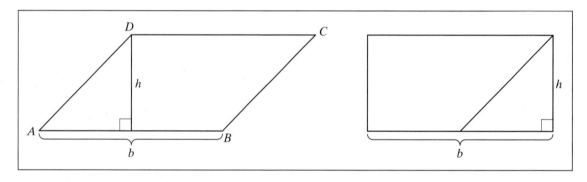

61. Area of a trapezoid is $\frac{1}{2}(b_1 + b_2)h$

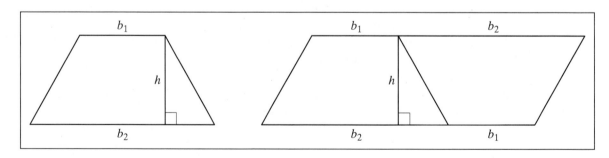

62. Area of a circle is πr^2.

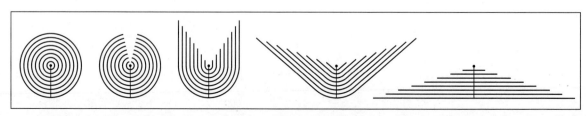

63. Create a semiregular tessellation out of at least two regular polygons. Your design should cover a minimum of an 8-inch-square section of paper.

64. a. Only three regular polygonal regions can be used to tessellate the plane by itself. Which shapes are these?
    b. Other tessellations are made of combinations of regular polygonal regions. Is it possible to tessellate the plane with nonregular figures? For example, can you tessellate the plane with any triangle? Any quadrilateral? Any pentagon? Any hexagon? Justify your answers.
    c. Make a general statement about nonregular figures that will tessellate the plane.
    d. So far we have only used convex polygonal regions. Is it possible to tessellate the plane using concave polygonal regions? What criteria must the shape meet to tessellate the plane?
    e. Draw at least one concave figure that will tessellate the plane, and then draw the tessellation. Will every quadrilateral tessellate the plane? Explain.
    f. Find a semiregular tessellation anywhere outside of the classroom. Sketch it, and identify the location of the tessellation.

65. Suppose you need to find the height of a tall tree but have no way of measuring it directly. Use the diagram below to write a complete explanation of how you could determine the tree's height.

Writing/Discussing

66. Why do you think that throughout history civilizations have developed measurement systems? Why do you think the measurement systems have changed over time?

67. Discuss any beliefs you have or may have had concerning the relationship(s) between perimeter and area. Has your thinking changed? If so, how?

68. When finding area of shapes on a geoboard, how could you use ideas of decomposition?

69. In Activity 10.5, how did you see ideas of multiple representations being used?

70. Choose one of the proofs of the Pythagorean theorem, and write up the proof so that someone could read and understand it without you being there to explain what you wrote.

Appendix A
Pages to Accompany Selected Activities

Appendix Outline:

        Page to Accompany Activity 6.4

        Page to Accompany Activity 6.8

        Page to Accompany Activity 10.11

        Page to Accompany Activity 10.12

Activity 6.4 *Appendix A* **475**

Use this centimeter strip for measuring.

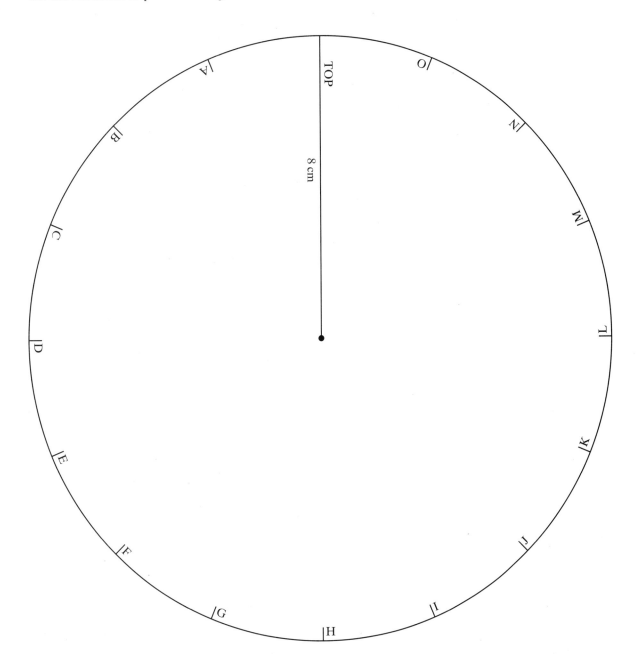

Activity 10.12 **Appendix A** **481**

Use this centimeter strip for measuring.

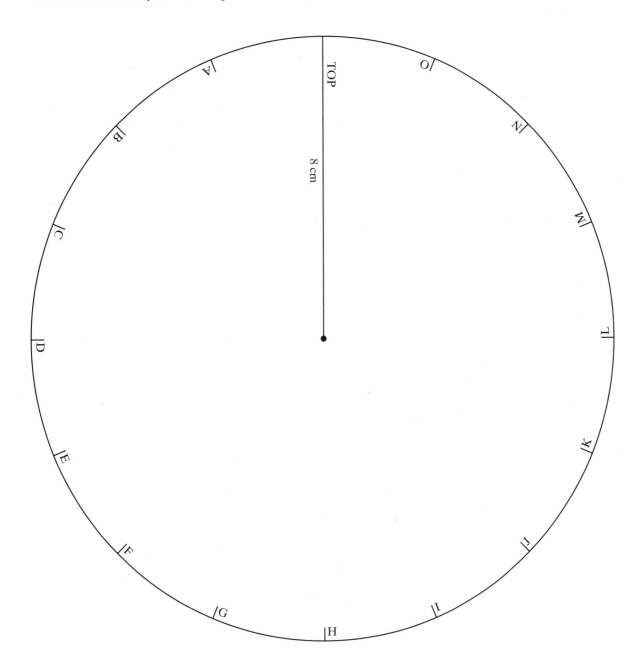

Solutions to Odd-numbered Exercises & More Problems

Chapter 1

1. Various solutions are possible; one possibility is: 8, 16, 32; another possibility is 7, 11, 16

3. 116, 159, 209

5. 16, 19, 22

7. Answers will vary.

9. a. The following table shows one way in which Rebekah could get 6 cups and 1 cup of water. It lists the amount of water at each step in each of the cups. Only one transfer has been made at each step. Remember that because it is water that is needed, Rebekah could afford to throw away the unused quantities of water.

5 cup vessel	8 cup vessel
0	8
5 (throw away)	3
3	0
3	8
5 (throw away)	6 → Amount needed for the recipe

    b. 5   1

    c. Once Rebekah gets 1 cup of water, she can get any other quantity (in whole cups) by storing the 1 cup in another vessel and adding on any more cups of water.

11. a. You can get all those and only those values that are sums or multiples of 5 and 8; that is, all those values that are of the form $8x + 5y$, where x and y are non-negative whole numbers. The numbers 7, 14, and 17 cannot be obtained. The following combinations show some possible amounts.

$18 = 8 \cdot 1 + 5 \cdot 2$ $31 = 8 \cdot 2 + 5 \cdot 3$

$41 = 8 \cdot 2 + 5 \cdot 5$ $43 = 8 \cdot 1 + 5 \cdot 7$

$52 = 8 \cdot 4 + 5 \cdot 4$

In fact, every number greater than 27 can be obtained.

    b. The numbers 1, 2, 3, 4, 6, 7, 9, 11, 12, 14, 17, 19, and 27 cannot be obtained.

13. a. $23.01.

    b. At the end of December; at the end of March the following year.

15. 9 minutes.
Adam can put the first two waffles into the griddle and let the first side of both cook—this takes 3 minutes. He then flips over one waffle but removes the second one from the griddle and puts in the third one. At the end of end of another 3 minutes, both sides of the first waffle and one side each of the second and the third waffles are done. So he removes the first waffle and lets the second side of both the second and third waffles cook for another 3 minutes. Thus, total time taken is $3 + 3 + 3 = 9$ minutes.

17. The first number is larger. It can be explained by using the Distributive Property and the formula for the sum of the first n numbers.

19. Pull out a ball from the box marked "Red Balls and White Balls." This should give us the solution. For example, if the ball you extract is white, then this box should be marked "White Balls," the one marked "White Balls" should be marked "Red Balls," and the one marked "Red Balls" should read "Red Balls and White Balls."

21. The order is: Tom—156 lb, Harry—152 lb, Max—138 lb, John—135 lb, Eric—130 lb, and Kevin—120 lb.

23. The plants could be arranged as follows:

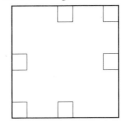

25. 28th day

27. Sara—$39, Cathy—$21, Tina—$12

29. 1

31. First move should be to the middle peg.

33. If you have an even number of disks, the first move should be to the peg that is not the target peg; if you have an odd number of disks, the first move should be to the target peg

35. You always get a number of the form *nnnnnn*, where *n* is the original number. Thus, with 3 you get 333333; with 5, 555555. This happens because $273 \cdot 407 = 111111$. You could factor 111111 differently to get other numbers in place of 273 and 407.

Chapter 2

1.

72	▼⟨▼▼	∩∩∩∩∩∩∩||	⋮⋮	LXXII
1,273	⟨⟨▼⟨▼▼▼	↑ഉ∩∩∩∩∩∩∩|||	⋮⋮⋮	MCCLXXIII
1,813	⟨⟨⟨⟨▼▼▼	↑ഉഉഉഉഉഉഉഉഉ∩|||	☰	MDCCCXIII
1,965	⟨⟨⟨▼▼⟨⟨⟨⟨▼▼▼▼▼	↑ഉഉഉഉഉഉഉഉഉ∩∩∩∩∩∩|||||	☰	MCMLXV
121	▼▼ ▼	ഉ ∩∩|	÷	CXXI
231	▼▼▼ ⟨⟨⟨⟨⟨▼	ഉഉ∩∩∩|	≡	CCXXXI

3. a. A numeration system is said to be a place-value system if the value of each digit in a number in the system is determined by its position in the number. The place value of a digit is a description of its position in a given number that determines the value of the digit. In the given number, 2 represents 200, 5 represents 50, 4 represents 4, 7 represents 7/10, and 1 represents 1/100.

    b. A numeration system is said to be multiplicative if each symbol in a number in that system represents a different multiple of the face value of that symbol. In the given number, the 2 represents the multiple $2 \cdot 10^2$, the 5 represents $5 \cdot 10^1$, the 4 represents $4 \cdot 1$, the 7 represents $7 \cdot 10^{-1}$, and the 1 represents $1 \cdot 10^{-2}$.

c. A numeration system is said to be additive if the value of the set of symbols representing a number is the sum of the values of the individual symbols. In the given number, the value of the number is equal to $200 + 50 + 4 + 0.7 + 0.01$.

d. A numeration system is a unique representation system if each numeral refers to one and only one number because there is only one number that is represented by 254.71.

5. Answers will be of the form $53x0$, where x represents any digit but could also have digits in the places higher than thousands.

7. a. 5,546   b. 206   c. 371   d. 244,107

9. a. x = five   b. x = seven

11. Starting with 20_{four} means you have 2 longs. Trade in one long for 4 units. Now, you have 1 long and 4 units. Removing 1 long and 3 units leaves you with one unit.

13. Base three: 1 cube, 2 longs, 2 units; base six: 5 longs, 5 units.

15. $3 \cdot 5^2 + 2 \cdot 5^1 + 4 \cdot 5^0 + 1 \cdot 5^{-1} + 3 \cdot 5^{-2} + 2 \cdot 5^{-3}$

17. a. i. 326,057   ii. 1,802,036
    b. i. ○ ○○ ○ ○ ○ ○○ ○○
         ○ ○ ○○ ○ ○○   ○
         ○ ○
       ii. ○ ○○ ○○ ○   ○   ○   ○
           ○   ○ ○ ○○ ○○ ○○
           ○ ○

19.
```
        31   41   43
   21   52  102  104
   34  105  115  121
   51  122  132  134
```

21. 700603_{eleven}

23. 3 gal 2 qt 1 cup; base 4

25. 2 yd 2 ft 8 in; no base

27. Trade and regroup as needed

29. 1 yd 3.5 in.

31. 1 qt 15 tbsp

33. Even though these situations do not have one consistent base, these conversions are similar to converting a number from one base to another because of the method of looking for the largest power of a unit (base), subtracting that amount, and repeating this process.

35. Using an organized list and looking for a pattern is one way to find the solution to the Sultan problem. The place of the first wife to receive a ring in the nth round is 3^{n-2} more than the place of the first wife to receive a ring in the $(n-1)$st round, where $n \geq 2$. Thus, the generalized solution is

$$1 + \Sigma\, 3^k, \text{ where } k = 0 \text{ to } n - 2.$$

The answer is 9842 wives. This number is $111111112_{\text{base three}}$. Writing the number in base three will help you better understand the problem and where the idea of multiples of 3 fits in.

37. Draw out one coin from the box labeled 35¢. If it is a dime, then the box should be labeled 20¢, and if it is a quarter, then it should be labeled 50¢. Given that all the three boxes are labeled incorrectly, it is now easy to work out the other labels. (This problem is similar to #19 in Chapter 1.)

39. 222222 is the largest six-digit base-three numeral because if you add 1, you would have 1000000 (a seven-digit numeral).

41. They were using base 7; the adventurer was using base 10.

43. Four weights—1, 2, 4, and 8 grams; six weights—1, 2, 4, 8, 16, and 32 grams; 63 grams

45. 8 is the larger number; 5 is the larger numeral

47. a. six weights—1, 2, 4, 8, 16, and 32 oz
    b. one weight—1 oz
    c. four weights—1, 3, 9, 27 oz
    d. base two, base three
    e. both use base-two system

Chapter 3

1. a. $(3 + 4) + 8 = 3 + (4 + 8)$
   b. $7 \cdot 5 = 5 \cdot 7$
   c. $4 + 0 = 4$
   d. $3 \cdot 1/3 = 1$

3. a. Yes, a is the additive identity because a added to any element will equal that element.
   b. Yes, order in adding does not change the result; the table is symmetric.
   c. Yes, b has c for an inverse because $b + c = a$.

5. $4 \div (8 - 4) \neq (4 \div 8) - (4 \div 4)$ because $1 \neq -1/2$.

7. Word problems will be similar to those in Activity 3.3

9. b, f, h, b, a

11. Illustrations will be similar to those in Activity 3.4.

13. a.

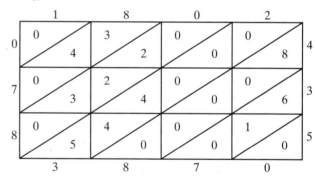

The product is 33,123.

b.

The product is 783,870.

15. a. $6\,{}^{13}\!1$ b. $2\,{}^{14}\,{}^{12}\,{}^{13}$
       $-2^3\,8^9\,7$ $-\cancel{X}\,{}^{2}\cancel{6}\,{}^{7}\cancel{8}\,{}^{6}7$
       $\overline{\ \ 3\ 4\ 4\ }$ $\overline{\ \ 7\ 6\ 6\ }$

17. △ + □ ○

19. △ □ ÷ ○

21. ○ − □ ÷ ◇

23. The following is one way to write each of the numbers. There may be other ways.

$1 = (4 \div 4) + (4 - 4)$
$2 = (4 \div 4) + (4 \div 4)$
$3 = (4 + 4 + 4) \div 4$
$4 = 4 + ((4 - 4) \cdot 4)$
$5 = (4 + (4 \cdot 4)) \div 4$
$6 = 4 + ((4 + 4) \div 4)$
$7 = (4 + 4) - (4 \div 4)$

$$8 = 4 + 4 + 4 - 4$$
$$9 = 4 + 4 + (4 \div 4)$$
$$10 = (44 - 4) \div 4$$

25. Word problems will be similar to those in Activities 3.4 and 3.5.

27. The set must be $\{1, 2, 3, 4, \ldots\}$ = natural numbers. Because the set is closed under addition and 1 is in the set, then $1 + 1 = 2$ is in the set. Then, $2 + 1 = 3$ is in the set, and so on.

29. $-x$ does not mean that $-x$ is negative; it simply means the opposite of x. If $x < 0$, then $|x| = -x$.

31. a. 234
            153
             42
            302
         + 510
         ———
         2125_{six}

b. 145
     34
    321
     40
   + 152
   ———
   1220_{six}

33.
```
              -3
          -3       0
       -2    -1       1
     3   -5    4    -3
   5   -2   -3    7   -10
```

35. 26 floors.
The answer can be found as follows:
$4 + 10 - 6 + 15 - 3 + 2 + 4 = 26$.

37. a. No, if 7 is removed, the set will not be closed under addition (e.g., $4 + 3 = 7$ does not belong to the set).
b. If 7 is removed, the set will still remain closed under multiplication because the only way to have a product of 7 is to multiply 7 and 1.
c. If 6 is removed from the set, the set will lose the property of closure under both addition and multiplication because $2 \cdot 3 = 6$.

39. a. The distributive property of multiplication over addition.
b. Consider the multiplication of 54 by 13 in the standard algorithm.

$$\begin{array}{r} 54 \\ \times\ 13 \\ \hline 162 = 150 + 12 = 50 \cdot 3 + 4 \cdot 3 \\ 540 = 500 + 40 = 50 \cdot 10 + 4 \cdot 10 \\ \hline 702 \end{array}$$

Thus, the process of the multiplication is the same as in standard multiplication, but the number of steps is increased and the multiplication is with the multiple of the nearest power of 10 in this case.
c. Answers will vary.
d. Multiplication by multiples of 10 and single digits makes the operation easier.

41. Her students would find $42 \div 6$ by repeatedly subtracting 6 and counting the number of times they were able to subtract 6 until they reached 0. Because division can be thought of as repeated subtraction, this algorithm works.

Chapter 4

1. a. $2 \cdot 3 \cdot 5 \cdot 7$ b. $2^3 \cdot 17$ c. $2 \cdot 3^3$
d. $2^3 \cdot 5^3$ e. $2^4 \cdot 5^3$

3. $110 = 2 \cdot 5 \cdot 11$ $204 = 2 \cdot 2 \cdot 3 \cdot 17$
Start by dividing numbers by 2 because they are both even; then continue dividing by factors until you reach a prime number.

5. 7, 29, 43, 57, and 143 are prime. The others are composite.

7. a. $12/28 = 3/7$; looked for greatest common factor for 12 and 28, which is 4, and then divided the numerator and denominator by 4
b. $90/105 = 6/7$; wrote prime factorization for each number, looked for greatest common factor, which is 15, and then divided the numerator and denominator by 15
c. $\dfrac{2 \cdot 5^2 \cdot 11}{(3 \cdot 7)}$; looked for common factors to divide out of numerator and denominator

9. 310 has more than 6 factors because its prime factorization is $2 \cdot 5 \cdot 31$. Thus, its factors are 1 and 310; 2 and 155; 5 and 62; 10 and 31.

11. 3 and 7

13. a. Yes, because it meets the requirements for divisibility by 3 *and* 8.
b. Yes, because it meets the requirements for divisibility by 5 *and* 8.
c. No, because it does not meet the requirements for divisibility by 9.
d. Yes, because it meets the requirements for divisibility by 4 *and* 9.

15. a. 44   3. a.   A   b.   3   c.   16   d.   8

17.
```
0  1  2  3
1  2  3  0
2  3  0  1
3  0  1  2
```
a. 0   b. 1   c. 2

19. a. 1   b. 3   c. 3   d. 2   e. 5   f. 2
g. 4   h. 7   i. 4   j. 0   k. 4   l. 0
m. 1   n. 3   o. 6   p. 4

21. a. 3 because $5 \cdot 3 = 15 = 1 \pmod 7$.
b. There is no multiplicative inverse for 6 (mod 9).
c. 3 because $3 \cdot 3 = 9 = 1 \pmod 4$.

23. p^{13} has 14 factors: $p^0, p^1, p^2, p^3, \ldots, p^{13}$.

25. No, one integer would have to be even and, therefore, not prime.

27. a. perfect   b. not perfect   c. not perfect
d. perfect

29. A number in the form p^4 has exactly five divisors. Some examples are 16, 81, 625.

31. a. 0, 2, 4, 6, and 8
b. 3 and 9
c. 5

33. GCD $(x^2, y^2) = 1$. Because x and y do not have any common factors other than 1, the products $x \cdot x$ and $y \cdot y$ will still not have any other common factors.

35. If a number is divisible by 3 and 4, it is divisible by 12.

37. a. false   b. true

39. 1, 2, 3, 4, 6, 8, 9, 12, 16, 18, 24, 36, 48, 72, and 144 rows

41. 600 days; least common multiple

43. N is greater than 1 because it is 1 more than the product of some whole numbers, so it must be positive also. None of 2, 3, 5, $\ldots$, p can divide N because it is one more than a multiple of each of these (because 1 was added to the product).

45. Because zero cannot be multiplied by any number to get a certain number n, zero is not a factor of any number. This is why division by zero is undefined.

47. Answers will vary.

49. 51 eggs

51. container 4

53. a. If every digit in the number appears exactly three times, then the sum of the digits of the number will be a multiple of 3 and hence will be divisible by 3.
b. We can make a similar conjecture for the number 9.

55. Writing a three-digit number and then repeating it to get a six-digit number is the same as multiplying the three-digit number by 1,001. Also, $7 \cdot 11 \cdot 13 = 1,001$. Hence when we divide successively by 7, 11, and 13, we actually divide the number by 1,001, thus getting the original number.

57. 23; Look at the multiples of 7 increased by 2, until you find the required number.

59. a. The arrangement was $(1 + 100) + (2 + 99) + (3 + 98) + \cdots + (50 + 51)$. The sum of every pair of numbers within the parenthesis is 101, and there are 50 such pairs (one-half of 100). Thus, the sum of the numbers is $50 \cdot 101 = 5050$.
b. We may try the above process with different sets of numbers. This will help to generalize the method. Thus, the general formula will be $(n/2)(n + 1)$.

61. a. Some numbers that have a large number of factors are 24 (eight factors), 30 (eight factors), 36 (nine factors), 40 (eight factors), 48 (ten factors)
b. Some numbers that have only two factors are 29, 31, 37, 41, 43, 47.

63. a. False; 16 is divisible by 4, but it is not divisible by 12.
b. True; because 4 is a factor of 12, any number that is divisible by 12 is also divisible by 4.

65. Mod seven and base seven are alike in that both use seven digits; in base seven the digits are 0–6, and in mod seven the digits are any seven consecutive digits, but usually we use 0–6 or 1–7 (for clock arithmetic). They are also alike in that, in both cases, grouping is done by sevens (although the grouping is different), and when a group of seven is obtained, the situation (counting, performing an operation) changes. In mod seven, this change takes the form of starting over at zero (or the first digit). In base seven, this change takes the form of regrouping groups of seven. Mod seven and base seven are different in that mod seven is a mathematical system with properties of a mathematical system. Base seven is a numeration system with properties of a numeration system.

67. Answers will vary; some possibilities are:
$4 - 5 = 5$, $2 - 5 = 3$; $0 - 3 = 3$; $3 \div 5 = 3$;
$2 \div 4 = 2$; $1 \div 5 = 5$.

69. The nth term of a triangular number is equal to the nth term of a square minus the $(n - 1)$st term of a triangular number; in other words

$$n(n + 1)/2 = n^2 - n(n - 1)/2$$

71. The pattern of squares is $1^2, 2^2, 3^2, \ldots$. At the nth stage, the number of squares will be n^2. There may be other patterns.

73. 33,552

75. c. The sum of the numbers in the circle is twice the number two rows directly below the number surrounded by the circled numbers.

77. Let x be the first number in a Fibonacci sequence. Then, the first 10 consecutive Fibonacci numbers are: $x, x, 2x, 3x, 5x, 8x, 13x, 21x, 34x, 55x$. The sum of these 10 numbers is $143x$ which is divisible by 11, so the sum of any 10 consecutive Fibonacci numbers is a multiple of 11.

Chapter 5

1. a. $6 \cdot 6 \cdot 6 = 216$ elements
b. The set of vowels considered is {a, e, i, o, u}. There are $26 \cdot 5 \cdot 26 = 3,380$ "words." If the letter "y" is considered as a vowel, which it sometimes is, there are $26 \cdot 6 \cdot 26 = 4056$ words. Note that this solution allows for vowels in the first and third positions of each word. If vowels are not allowed in these positions, there are either $21 \cdot 5 \cdot 21 = 2,205$, or $21 \cdot 6 \cdot 21 = 2,646$ words.
c. $52 \cdot 51 = 2,652$ possible sets of 2 cards.
d. $2 \cdot 2 \cdot 2 \cdot 2 \cdot 2 = 32$ possibilities.

3. For five sandwiches and four drinks, there are $4 \cdot 5 = 20$ sandwich-drink combinations we can choose. If there are an additional three desserts, then, there are $4 \cdot 5 \cdot 3 = 60$ combinations of sandwich-drink-desserts from which to choose.

5. The sample space for the experiment is as follows:

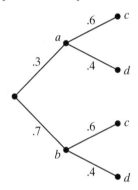

a. $P(a \& d) = 0.3 \cdot 0.4 = 0.12$.
b. $P(a \text{ or } d) = P(a \& c) + P(a \& d) + P(b \& d) = 0.3 \cdot 0.6 + 0.3 \cdot 0.4 + 0.7 \cdot 0.4 = 0.58$.

7. There is a total of $(2)^4 = 16$ possible outcomes. The subset of the sample space that is favorable to an outcome of *at least* 3 heads is E, where E = {HHHT, HHTH, HTHH, THHH, HHHH}. Hence, P(at least 3 heads) = 5/16.

9. If we assume that all the faces of the die are distinguishable from each other (for example, have different colors), then the number of different ways to mark the die is $6 \times 4 \times 2 = 48$.

11. $10 \times 9 \times 8 \times \ldots \times 3 \times 2 \times 1 = 3,628,800$

13. a. $9,000 (about 75% of $12,000)
b. $15,000. The value is $5,000 (about 25% of $20,000), so the car has depreciated by $15,000.
c. $7,000 (about 35% of $20,000)
d. Dani should do it *within two years!* (By the time the car is two years old, it has depreciated by more than 50%).

15. According to the table, we must sum 5.5%, 10.0%, 12.8%, and 19.4%. So, the answer is 47.7%.

17. In 1970, 21.3% of the population had taken one or more years of college. This percent is the sum of 10.6% and 10.7%. 10.7/21.3 = 50.2%, which is the percent of people who had begun college and who had finished a four-year degree.

19. Post secondary education has steadily increased since 1970, and the percent of people who have not completed high school has steadily decreased.

21. In this problem, unlike the situation in #20, order does not matter, so we get $(31 \cdot 30)/2 = 465$ double scoops are possible.

23. In bowling, the bowler can knock down 0–10 pins on the first roll. The table at the top of page 487 summarizes the possibilities:

No. of Pins Knocked Down	No. of Ways
0	1
1	10
2	45
3	120
4	210
5	252
6	210
7	120
8	45
9	10
10	1
Total	1,024

25. Answers will vary. Here is one example: Two cards are drawn from a standard deck, and the first card is not replaced before drawing the second card. Find the probability of drawing two red cards.

P(red and red) = 26/52 · 25/51 = 650/2652 = (about) 24.5%.

27. Although the probability of rolling an even number is equal to the probability of rolling an odd number, the game is unfair because it is likely that you will win more than lose. The expected winnings from the game are equal to

(−$2) × 1/6 + ($4) × 1/6 + (−$6) × 1/6 + ($8) × 1/6 + (−$10) × 1/6 + ($12) × 1/6 = $1.

29. Expected number of accidents = 0(.62) + 1(.15) + 2(.10) + 3(.08) + 4(.02) + 5(.03) = .82

31. a. If the y-axis started at 0 then the distortion factor would be undefined in so far as division by 0 and division of infinity by infinity are undefined.
   b. Distortion factor of 1 means that the graph's maximum change is proportional to the data's maximum change (i. e., that the graph represents the real data adequately in terms of ratio (although not in terms of magnitude).
   c. In order to eliminate any distortion, the graph should be proportionally squeezed or stretched in horizontal directions until the distortion factor is equal to 1.

33. Suppose p denotes the probability that a woman wins. Then P(Lois) = P(Kathy) = P(Diana) = p, and P(Paul) = P(Norm) = $2p$, since each man is twice as likely to win as any woman. Now the sum of these probability must be 1. That is, $p + p + p + 2p + 2p = 1$. So, $p = 1/7$.
   a. The probability that a woman wins is P(any woman wins) = 1/7 + 1/7 + 1/7 = 3/7.
   b. The probability that either Paul or Diana wins the tournament is P(Paul or Diana) = P(Paul) + P(Diana) = 2/7 + 1/7 = 3/7.

35. a. There are 30 individuals altogether. Hence, 10 people can be selected from a group of 30 people in (30 · 29 · 28 · 27 · 26 · 25 · 24 · 23 · 22 · 21) ÷ (10 · 9 · 8 · 7 · 6 · 5 · 4 · 3 · 2 · 1) = 30,045,015 ways.
   b. The 5 smokers can be chosen in (20 · 19 · 18 · 17 · 16) ÷ (5 · 4 · 3 · 2 · 1) = 15,504 ways. The nonsmokers can be selected in (10 · 9 · 8 · 7 · 6) ÷ (5 · 4 · 3 · 2 · 1) = 252 ways. Hence, there 15,504 · 252 = 3,907,008 ways in which this can be done!

37. Sample Space = {1, 2, 3, 4, 5, 6, 8, 10, 12, 9, 15, 18, 16, 20, 24, 25, 30, 36}; Probability distribution: P(1) = P(36) = P(9) = P(16) = P(25) = P(36) = 1/36; P(2) = P(3) = P(5) = P(8) = P(10) = P(15) = P(18) = P(20) = P(24) = P(30) = 2/36 = 1/18; P(4) = 3/36 = 1/12; P(6) = 4/36 = 1/9 = P(12).

39.

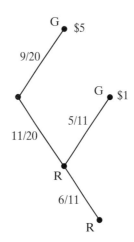

A: sets alarm   F: does not wake in time
B: does not set alarm G: wakes in time
C: alarm rings   H: does not wake in time
D: alarm does not ring J: wakes in time
E: wakes in time   K: does not wake in time

b. From the above tree, the probability that Jamal wakes in time for his 8:30 class is P(in time for class) = 0.8 · 0.9 · 0.8 + 0.8 · 0.1 · 0.3 + .2 · .1 = 0.66.

41. a.

c. Expected earnings = $5 · 9/20 + $1 · (11/20) · (5/11) = $2.25 + $0.25 = $2.50

43. a. There are 2^8 = 256 possible outcomes.
   b. There are 56 ways to have exactly 3 heads.
   c. 247 outcomes have at least 2 heads.
   d. 126 outcomes have 4 or 5 heads.
   e. The entire sample space could be listed in a systematic manner and then inspected to see how many of them contain 2 or more heads. Another, more efficient way to determine how many outcomes have at

least 2 heads is to determine how many outcomes have fewer than 2 heads and then subtract the result from 256.

45. a. 9/60 = 3/20
    b. 12/60 or 1/5
47. a. Both players have the same chance of winning because both can win in 2 ways of the 4 possible outcomes.
    b. Yes, the game is fair because the two players have the same likelihood of winning.
49. Let b = boy and g = girl.
    The sample space S is: S = {bbb, bgg, bgb, bbg, gbb, ggb, gbg, ggg}
    Favorable outcomes are bgg, bgb, bbg, gbb, ggb, gbg, ggg. Hence, the probability is 7/8.
    Alternatively, observe that P(at least one girl) = 1 − P (all 3 children are boys)
    But P(all 3 children are boys) = 1/8. Hence, P(at least one girl) = 1 − 1/8 = 7/8.
    You must assume that the probability of a boy being born is equal to the probability of a girl being born.
51. 45 matches
53. 45 handshakes
55. answers will vary
57. Various interpretations are possible. For example, one could compare Communist bloc (those that existed in 1988) countries with other countries and conclude that they won far more medals that non-Communist countries. The countries could be grouped by continent (viz., Africa, Asia, Australia, Europe, North America, South America) and draw conclusions about medal distribution (e.g., no South American countries were among the top 20 medal winners.)
59. The number of medals won by the USSR, East Germany, and the USA are outliers using the definition given.
61. In this particular case, the median probably best represents the average number of medals won by the top 20 countries—19 or 20. This median almost exactly coincides with the mean number of medals won by the 17 non-outlier countries, which is 19.8. It would be a misrepresentation to eliminate the outliers by using the median of the other 17 countries. An even better indicator of the average number of medals won is probably the interquartile range because 55% of the values fall within this range.
63. Answers will vary depending on personal predictions made and how close these predictions were to the theoretical probabilities. A reasonable definition of the probability of an event is as follows:
    If an event has n equally likely outcomes and its sample space has m equally likely outcomes, then the probability of the event is n/m.
65. An *outcome* of an experiment is the result of performing the experiment. The *sample space* of an experiment is the set of all possible outcomes of the experiment. An *event* in an experiment is a subset of the sample space of the experiment. The event is said to have occurred if the outcome of an experiment corresponds to one of the elements of the event.
67. P(E) is always less than or equal to 1 because the number of outcomes in an event must be less than or equal to the number of outcomes in the sample space. P(E) is always greater than or equal to 0 because $P(E) = n/m$ and $m > 0, n \geq 0$.
69. answers will vary

Chapter 6

1. Team.
3. Math is fun.
5. Answers will vary.
7. a. 1/16; (answers will vary but could include:
      $1/4 \cdot 1/4$; $1/8 \cdot 1/2$; $1/12 \cdot 3/4$)
      1/12; (answers will vary but could include:
      $1/2 \cdot 1/6$; $1/4 \cdot 1/3$; $1/6 \cdot 1/2$)
      1/4; (answers will vary but could include:
      $1/2 \cdot 1/2$; $1/3 \cdot 3/4$)
      1/8; (answers will vary but could include:
      $1/2 \cdot 1/4$; $1/4 \cdot 1/2$; $1/6 \cdot 3/4$)
   b. Answers will vary.
9. Answers will vary; some possibilities include:
   a. 1/3, 2/6   b. −6/10, −9/15   c. 0/1, 0/5
11. a. 78/77   b. 17/91   c. −17/45
13. a. 35/78   b. −32/63   c. 34/63
15. Answers may vary.
    a. 15/18, 31/36   b. −17/28, −35/56
    c. 144/209, 145/209
17. Place appropriately in this order: 0.75 (b), 1.125 (c), 2 (d), 2.25 (a)
19. Divide the paper into tenths, and shade two tenths. Show by folding, for example, that this is one-third of six tenths.
21. It would need to have a denominator with values less than or equal to the other.
23. Illustrations will vary; 9/10.
25. 14
27. Charge $3.33 for 576 square inches of material.
29. 18 2/21 pizzas.
31. When n is a multiple of 2.
33. 15 bookmarks; 3/4 inch left over.
35. a. a < c   b. b > d
    c. Answers will vary; a/b = c/d.
37. 5/6 = 1/2 + 1/3   13/12 = 1/3 + 3/4
    13/36 = 1/4 + 1/9   26/21 = 2/3 + 4/7
39. There are infinitely many answers. Some possibilities are:
    a. 1/3, 4/5, 15/2
    b. 2/3, 3/10, 1/10
    c. 3/4, 1/2, 4/3
41. 2/3 ÷ 2/5 is asking "How many 2/5 are in 2/3?"; because there are 5/2 2/5s in 1 unit, then we must multiply 2/3 by 5/2 to find out how many 2/5 are in 2/3; illustration should demonstrate why there are 5/2 2/5s in one unit.

Chapter 7

1. Answers may vary; a block with 2 out of the 10 units shaded (or 20 out of 100 units, etc.).
3. a. 12.084   b. 1.625   c. 6
5. a. 3.2, 3.$\overline{22}$, 3.23, 3.2$\overline{3}$, 3.2$\overline{3}$
   b. −1.4$\overline{54}$, −1.$\overline{454}$, −1.$\overline{454}$, −1.454, −1.45
7. Write the decimals out to several more places, and then compare.
9. Answers will vary; some possibilities are −2.295, −2.297.

11. e
13. a. 0.875   b. 0.$\overline{3}$ c. 0.9$\overline{6}$
15. a. 13/30   b. 35/99
17. Answers will vary; some possibilities are $\frac{\pi}{4}$ and $\frac{\sqrt{3}}{2}$.
19. Real Numbers

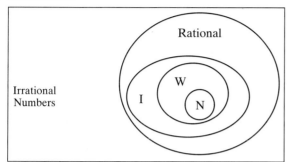

I = set of integers, W = set of whole numbers, N = set of natural numbers.

21. 3/5
23. The ratio of boys to girls will become smaller.
25. You must consider that values exist continuously to the right of the repeated value.
27. Only after performing the multiplication or division is it essential to determine where the decimal place should be located.
29. a. 88.6; 5 · 17 is 85   b. 15.6; 68 ÷ 4 is 17
    c. 14.6; 6 · 17 is 102, 102 · 0.1 is 10.2
    d. 0.132; 16 · 8 is a three-digit number; then divided by 1,000, it would be a three-digit decimal.
    e. 4.71; 0.15 goes into 0.7 about 4 times.
    f. 4.68; 400 ÷ 80 is 5
    g. 2.2; 0.09 goes into 0.198 about 2 times.
    h. 0.7615; 0.9 goes into 0.7 about 0.7 times
    i. 2.096641; 5 · 0.3 is 1.5
31. A fraction can be written as a terminating decimal if it can be expressed with a denominator that is a power of 10. But 7, having prime factors other than 2 and/or 5, is not a factor of any power of 10.
33. $11\frac{1}{9}\%, 22\frac{2}{9}\%, 66\frac{2}{3}\%$.
35. Yes, each side of the square could be $\sqrt{11}$ cm, which is possible to construct. Construct right triangles with these sides: 1, 1, $\sqrt{2}$ and 1, 2, $\sqrt{5}$. Then using these hypotenuses ($\sqrt{2}$ and $\sqrt{5}$) as legs, construct a new triangle having a hypotenuse of length $\sqrt{7}$. Use this segment (of length $\sqrt{7}$ cm) and a segment of length 2 cm to construct another right triangle; this triangle will have a hypotenuse of length $\sqrt{11}$ cm. Use this length to construct a square with sides of length $\sqrt{11}$ cm, and the area will be 11 cm².
37. a. 4%
    b. 32%
    c. 64%
39. Answers will vary; $\sqrt{2}$, 0.121221222..., $\sqrt{5}$.

Chapter 8

1. 0.01 billion per year between 1900 and 1910; 0.04 billion per year between 1900 and 1990.
3. Answers may vary. One estimation is 6.1 billion people, as this is consistent with the prior two decades' increase of 0.8 billion people. (Another could be 6.2 billion.)
7. Compare the slopes of the lines connecting the beginning and ending of each time period.
9. One reasonable answer is 273,000,000 people in 2000, as trends from the past few decades denote increases of just under 25,000,000 per decade.
11. Between 1930 and 1940. This was the period of the Great Depression, during which there was widespread unemployment.
13.

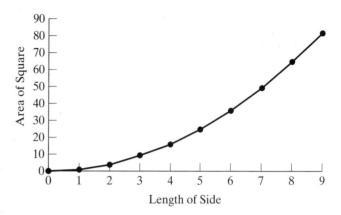

Both graphs curve upward rather rapidly. The circle graph increases much faster than the square graph (actually, π times faster).

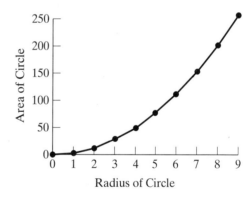

Both graphs curve upward rather rapidly. The circle graph increases much faster than the square graph.

15.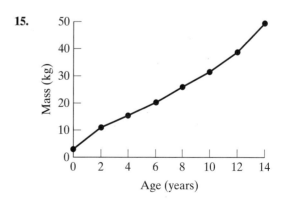

17. The largest product is 625. The smallest product is 49.
19. $3^{20} \cdot 2^{20}$ or $(3 \cdot 2)^{20} = 6^{20}$
21. 500,000 copies will be required for the cost of the options to be equal. Less, and the $17,500 copier will be cheaper.

More, and the $20,000 copier will be. Costs will be equal when $20,000 + 0.02x = 17,500 + 0.025x$; $x = 500,000$.

23. Area = $4x^2 + 100x$.

25. Let x represent the number of people making the trip and $C(x)$ represent the size of each member's contribution. The size of the contribution will decrease as the number of people making the trip increases. $C(x) = \frac{180}{x}$, $x > 0$.

For #27 and #29, the requested responses are underlined.

27.
Input	Output
1	18
2	21
3	24
9	42
x	$3x + 15$
$a/3 - 5$	a

29.
Input	Output
14	17
−20	0
20	20
230	125
x	$x/2 + 10$
$2a - 20$	a

31.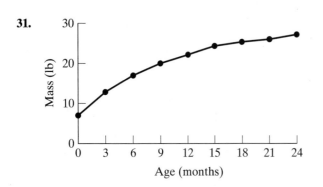

The weight of the average American baby increases over time. According to these data, the amount of increase decreases over time. The variables could be said to be positively related.

33.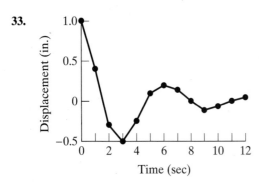

The pattern of the graph fluctuates from decreasing to increasing over time. The size of the fluctuation from zero decreases over time.

35.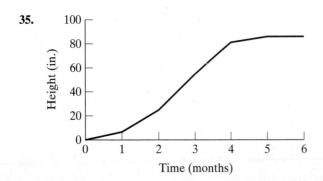

37.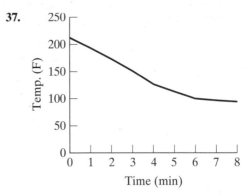

39. a. 1, 2, 4, 7, <u>11</u> For any number in the sequence, sum the number and its position in the sequence to get the next number in the sequence. (e.g., 4 is 3rd in the sequence, $4 + 3 = 7$)

b. 1, 2, 4, 7, <u>12</u> Starting with the third digit, sum the two previous digits and 1 to get the next digit. (e.g., For the fourth digit, the second and third (2 and 4) total 6, plus 1 equals 7)

c. 1, 2, 4, 7, <u>13</u> Starting with the fourth digit, add the previous three to get the next digit. (e.g., For the fourth digit, sum the first three (1 + 2 + 4) to get 7.)

41. 77°F to 104°F corresponds to 25°C to 40°C.

43. −40°C = −40°F.

45. This approach would work for approximations, but although there is a little more than 6% error for freezing (0°C), there is more than 11% error for 1,000°C. The accuracy required determines the appropriateness of the estimation.

47. Things to consider are the fact that the carrier earns $1 per paper and makes an additional $20, no matter how many papers are delivered. The more one is willing to work, if available, the more money one can earn.

49. a. $10
b. $P(x) = 10x - 250$, where $P(x)$ is the profit earned, and x is the number of lawns mowed.
c. 125 lawns
d. $P(x) = 10x - 195$
e. 120 lawns

51. $(176 - 88)/(1984 - 1960) = 3\ 2/3$ tons per year.

Chapter 9

1. a. 6 b. 10

3. Three points always lie in the same plane (are coplanar), but four points may not be coplanar.

5. a. Yes b. Yes, as long as the obtuse angle is between the two congruent sides
c. Yes d. No

7. 50° and 40°, respectively

9. Constructions

11. a. All three angles must be acute, and two sides must be congruent.
b. One angle must be obtuse, and two sides must be congruent.
c. One angle must be a right angle, and two sides must be congruent.
d. All three sides must be a different length. Median must connect one vertex with the midpoint of the opposite side.
e. All three sides must be congruent. Altitude is a segment connecting one vertex to the opposite side, forming a right angle.

13. a. The length of a side in each square; because we already know they are squares, we know their angles are congruent, and we need to determine if their side lengths are congruent.
  b. The length of two adjacent sides in each rectangle; because we already know they are rectangles, we know their angles are congruent, and we need to determine if their side lengths are congruent.
  c. The length of two adjacent sides and the measure of the included angle in each parallelogram; because we already know they are parallelograms, we know the opposite sides are parallel, and we need to determine if the opposite sides and opposite angles are congruent.
15. a. false   b. true   c. false   d. false   e. true
17. a and d are similar to a 5×8 rectangle.
19. Answers may vary, but some possibilities are:
  a. opposite sides parallel and one right angle
  b. opposite sides parallel and two consecutive sides congruent
  c. opposite sides parallel, one right angle, and two consecutive sides congruent
21. $a = (s - 2)150$
23.

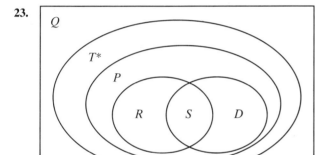

* with trapezoid defined as a quadrilateral with at least one pair of opposite sides parallel.

25. a. 70°   b. 60°   c. 60°   d. 110°   e. 70°
27. 135°; there are six triangles that can be formed in an octagon, so the sum of the measures of the angles is 1080°; because the octagon is regular, all eight angles are congruent; thus, they must each be 135°.
29. Yes, if the line lies in the plane.
31. Both terms refer to points that lie together; for collinear points, they lie on the same line; for coplanar points, they lie on the same plane.
33. Yes; the only place that the line can be perpendicular to both coplanar lines is at their point of intersection. Therefore, to be perpendicular to both of these lines means that the line must also be perpendicular to the plane.
35. This is not a good definition because the definition must include the fact that adjacent sides are congruent.
37. No, two congruent polygons means that all corresponding sides and angles are congruent, and thus all corresponding sides are in the ratio 1 : 1.
39. The included angle or the remaining side.
41. a. If $a > b > c$, then m∠A∠m∠B∠m∠C.
  b. If $a = b < c$, then m∠A = m∠B∠m∠C.
  c. If m∠B = m∠C > m∠A, then $b = c > a$.
43. a. Sometimes   b. True if an isosceles trapezoid; false if not
45. Yes, if all three pairs of corresponding sides are proportional, the triangles are similar.
47. No, rectangles always have their angles congruent, but the sides could be in different ratios.
49. a. $4/6 = 3/4.5 = 2/3$ because all are in the ratio of 1 : 1.5.
  c. yes
51. $x = 20/3$, y cannot be determined; $x = 6$.
53. Form an equilateral triangle from two 30°–60°–90° triangles. All angles will be 60° and all sides congruent. The measure of one leg of a smaller triangle will be one-half the measure of any side of the equilateral triangle, and hence the side opposite the 30° angle is half as long as the hypotenuse in one of the original triangles.
55. a. The base angles of an isosceles triangle are congruent.
  b. Use the angle bisector of the vertex angle.
57. Rebekah is correct. If two angles in one triangle are congruent to two angles in another, then the third angle must also be congruent, and then the triangles in this case are congruent by ASA.
59. Construct angle BDC and bisect it with ray DA. Draw segments AB and AC so that they are perpendicular at B and C. Triangles ABD and ACD are congruent (right triangle SA congruency). Therefore, segment AB is congruent to segment AC by CPCTC. Thus point A is equidistant from the sides of angle BDC.
61. a. The two base angles in each triangle are 45°; the hypotenuse in $\triangle ABC$ is $4 \div 2$ cm, and the hypotenuse in $\triangle DEF$ is $8\sqrt{2}$ cm.
  b. The triangles are similar by AA similarity.
  c. The triangles are not congruent because their corresponding sides are not congruent.
  d. Area $\triangle ABC = 8$ square cm; area $\triangle DEF = 32$ square cm.
63. 180°
65. *trapezoid*—quadrilateral with at least one pair of opposite sides parallel; *parallelogram*—quadrilateral with both pairs of opposite sides parallel; *rhombus*—parallelogram with one pair of congruent adjacent sides; *rectangle*—parallelogram with one right angle; *square*—parallelogram with one pair of congruent adjacent sides and one right angle; *kite*—quadrilateral with two pairs of congruent adjacent sides; *isosceles trapezoid*—trapezoid with at least one pair of congruent opposite sides
67. The farther down you go on the tree diagram, the more symmetry there is.
69. a. Between 26 and 72 miles apart
  b. You can determine the range by finding the difference (for the minimum distance) and the sum (for the maximum distance).
71. a. The new triangle is bigger, the ratio of the sides is 3 : 1.
  b. The new triangle is smaller, the ratio of the sides is $1/2 : 1$.
  c. The new triangle is similar to the original triangle.
73. It will be a rhombus; draw the diagonals in the rectangle, and then connect the midpoints as specified; we have that the quadrilateral formed is a parallelogram; furthermore, because the segments connecting midpoints are half the length of the diagonals, and because the diagonals are congruent in a rectangle, all four segments are congruent; thus, the quadrilateral is a rhombus; we cannot prove that the angles of $EFGH$ are right angles, so the most we can say is that the figure is a rhombus.

Chapter 10

1. a. 200 square inches   b. 60 inches
3. 1 yd × 24 yd;   2 yd × 12 yd;   3 yd × 8 yd;   4 yd × 6 yd

5. 8 square feet
7. 45.5 square units
9. 15 cm
11. 4 : 1
13. a. no   b. yes   c. no
15. 56.25π cm^2
17. The better bargain is the 5.5-cm-radius orange because its cost per cubic cm is less than the other orange.
19. 1. b   2. d   3. f   4. e
    5. g   6. c   7. a
21. a and c produce the same result.
23. Answers will vary; some possibilities are: BOOK for horizontal line, MOM for vertical line.
25. a. horizontal line through two opposite vertices
    b. horizontal and vertical lines, splitting rectangle in half
    c. horizontal and vertical lines, splitting figure in half
27. a. 4   b. 2   c. 3   d. 2
29. Find the area of the field (5,600 m^2), and take the square root; a square field with the same area should have a side length of approximately 74.83 m
31. Area of A = 8 square units; Area of B = 15 square units.
33. If you traveled 24 hours per day, the average speed in miles per hour would be about 13.1 mph. (*Note*: This is obtained from noting that you would travel about 314 miles per day.)
35. a. Area of the square is 144 in^2., perimeter of the square is 48 in., area of each semicircle is 18π in^2., circumference of each semicircle is 6π in.
    b. Area of the large circle is 36π cm^2, circumference of the large circle is 12π cm, the area of each small circle is 9π cm^2, the circumference of each small circle is 6π cm.
37. Yes, all *convex* quadrilaterals tessellate the plane. Each vertex point of the tessellation must be composed of each of the angles of the quadrilateral so that the sum of the angles at each vertex point is 360°.
39. Equilateral triangle, square, regular hexagon.
41. $12.50
43. Yes, the wire would be approximately 3.2 meters above Earth.
45. A 2 × 5 rectangle has perimeter of 14 units and area of 10 square units; a 6 × 6 rectangle has perimeter of 24 units and area of 36 square units.
47. 9 : 1
49. a. No, it is quadrupled.   b. It is quadrupled.
51. a. 4 : 1   b. 2 : 1   c. 4 : 1   d. 4 : 1
53. Divide the shape into figures that you know the areas of, and then sum the individual areas.
55. 134 square feet

57. Finding *all* the noncongruent polygons with five boundary pins and two interior pins is a challenging problem. A couple of examples of these "5–2" geoboard polygons are given below. (*Note*: Polygons A and B are not congruent, but both have five boundary pins and two interior pins.) You might want to find a systematic way to list all possible polygons (so you don't duplicate any or miss any) over the next few weeks and compare your list with other students in the class. One possible way to be systematic is to look first only at triangles, then only at quadrilaterals, and so on.

59.

61. Rotate a copy of the trapezoid into place as shown. A parallelogram is formed because both pairs of opposite sides are parallel (bases were already parallel, and angles on non-base sides are supplementary). Thus the area of a trapezoid is $1/2(b_1 + b_2)h$ (1/2 the area of a parallelogram formed by two congruent trapezoids).
63. Answers will vary.
65. Form similar triangles, and set up a proportion using corresponding sides.